Unmanned Aircraft Systems

Kimon P. Valavanis · Paul Y. Oh · Les A. Piegl

Unmanned Aircraft Systems

International Symposium on Unmanned Aerial Vehicles, UAV '08

Previously published in the Journal of Intelligent & Robotic Systems
Volume 54, Issues 1–3, 2009

 Springer

Kimon P. Valavanis
Department of Electrical and
Computer Engineering
School of Engineering and Computer Science
University of Denver
Denver, CO 80208
USA
kimon.valavanis@du.edu

Les Piegl
Department of Computer Science & Engineering
University of South Florida
4202 E. Fowler Ave.
Tampa FL 33620
USA
piegl@csee.usf.edu

Paul Oh
Applied Engineering Technology
Drexel University
3001 One Drexel Plaza
Market St., Philadelphia PA 19104
USA
paul@coe.drexel.edu

ISBN- 978-90-481-8076-9   e-ISBN- 978-1-4020-9137-7

Printed on acid-free paper.

9 8 7 6 5 4 3 2 1

springer.com

Contents

UAS NAVIGATION AND CONTROL

UAS SIMULATION TESTBEDS AND FRAMEWORKS

**Testing Unmanned Aerial Vehicle Missions in a Scaled
Environment** **297**
   K. Sevcik and P. Y. Oh

**A Framework for Simulation and Testing of UAVs in
Cooperative Scenarios** **307**
   A. Mancini, A. Cesetti, A. Iaulè, E. Frontoni, P. Zingaretti, and
   S. Longhi

UAS APPLICATIONS

Guest Editorial for the Special Volume On Unmanned Aircraft Systems (UAS)

Kimon P. Valavanis · Paul Oh · Les Piegl

Originally published in the Journal of Intelligent and Robotic Systems, Volume 54, Nos 1–3, 1–2.
© Springer Science + Business Media B.V. 2008

Dear colleagues,

This special volume includes reprints and enlarged versions of papers presented in the *International Symposium on Unmanned Aerial Vehicles*, which took place in Orlando FL, June 23–25.

The main objective of UAV'08 was to bring together different groups of qualified representatives from academia, industry, the private sector, government agencies like the Federal Aviation Administration, the Department of Homeland Security, the Department of Defense, the Armed Forces, funding agencies, state and local authorities to discuss the current state of unmanned aircraft systems (UAS) advances, the anticipated roadmap to their full utilization in military and civilian domains, but also present current obstacles, barriers, bottlenecks and limitations to flying autonomously in civilian space. Of paramount importance was to define needed steps to integrate UAS into the National Airspace System (NAS). Therefore, UAS risk analysis assessment, safety, airworthiness, definition of target levels of safety, desired fatality rates and certification issues were central to the Symposium objectives.

Symposium topics included, among others:

AS Airworthiness
UAS Risk Analysis
UAS Desired Levels of Safety
UAS Certification
UAS Operation
UAS See-and-avoid Systems
UAS Levels of Autonomy

K. P. Valavanis (✉) · P. Oh · L. Piegl
Department of Electrical and Computer Engineering,
School of Engineering and Computer Science, University of Denver,
Denver, CO 80208, USA
e-mail: kvalavan@du.edu, kimon.valavanis@du.edu

K. P. Valavanis et al. (eds.), *Unmanned Aircraft Systems*. DOI: 10.1007/978-1-4020-9137-7_1

UAS Perspectives and their Integration in to the NAS
UAS On-board systems
UAS Fail-Safe Emergency Landing Systems
Micro Unmanned Vehicles
Fixed Wing and Rotorcraft UAS
UAS Range and Endurance
UAS Swarms
Multi-UAS coordination and cooperation
Regulations and Procedures

It is expected that this event will be an annual meeting, and as such, through this special volume, we invite everybody to visit http://www.uavconferences.com for details. The 2009 Symposium will be in Reno, NV, USA.

We want to thank all authors who contributed to this volume, the reviewers and the participants. Last, but not least, The Springer people who have been so professional, friendly and supportive of our recommendations. In alphabetical order, thank you Anneke, Joey, Gabriela and Nathalie. It has been a pleasure working with you.

We hope you enjoy the issue.

Development of an Unmanned Aerial Vehicle Piloting System with Integrated Motion Cueing for Training and Pilot Evaluation

James T. Hing · Paul Y. Oh

Originally published in the Journal of Intelligent and Robotic Systems, Volume 54, Nos 1–3, 3–19.
© Springer Science + Business Media B.V. 2008

Abstract UAV accidents have been steadily rising as demand and use of these vehicles increases. A critical examination of UAV accidents reveals that human error is a major cause. Advanced autonomous systems capable of eliminating the need for human piloting are still many years from implementation. There are also many potential applications of UAVs in near Earth environments that would require a human pilot's awareness and ability to adapt. This suggests a need to improve the remote piloting of UAVs. This paper explores the use of motion platforms to augment pilot performance and the use of a simulator system to asses UAV pilot skill. The approach follows studies on human factors performance and cognitive loading. The resulting design serves as a test bed to study UAV pilot performance, create training programs, and ultimately a platform to decrease UAV accidents.

Keywords Unmanned aerial vehicle · Motion cueing · UAV safety · UAV accidents

1 Introduction

One documented civilian fatality has occurred due to a military UAV accident (non-US related) [1] and the number of near-mishaps has been steadily rising. In April 2006, a civilian version of the predator UAV crashed on the Arizona–Mexico border within a few hundred meters of a small town. In January 2006, a Los Angeles County Sheriff lost control of a UAV which then nose-dived into a neighborhood. In our own experiences over the past six years with UAVs, crashes are not uncommon. As Fig. 1 illustrates, UAV accidents are much more common than other aircraft and are increasing [2]. As such, the urgent and important issue is to design systems

J. T. Hing (✉) · P. Y. Oh
Drexel Autonomous Systems Laboratory (DASL),
Drexel University, Philadelphia, PA 19104, USA
e-mail: jth23@drexel.edu

P. Y. Oh
e-mail: paul@coe.drexel.edu

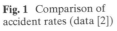
Fig. 1 Comparison of
accident rates (data [2])

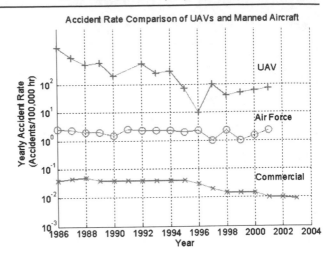

and protocols that can prevent UAV accidents, better train UAV operators, and augment pilot performance. Accident reconstruction experts have observed that UAV pilots often make unnecessarily high-risk maneuvers. Such maneuvers often induce high stresses on the aircraft, accelerating wear-and-tear and even causing crashes. Traditional pilots often fly by "feel", reacting to acceleration forces while maneuvering the aircraft. When pilots perceive these forces as being too high, they often ease off the controls to fly more smoothly. The authors believe that giving the UAV pilot motion cues will enhance operator performance. By virtually immersing the operator into the UAV cockpit, the pilot will react quicker with increased control precision. This is supported by previous research conducted on the effectiveness of motion cueing in flight simulators and trainers for pilots of manned aircraft, both fixed wing and rotorcraft [3–5]. In this present study, a novel method for UAV training, piloting, and accident evaluation is proposed. The aim is to have a system that improves pilot control of the UAV and in turn decrease the potential for UAV accidents. The setup will also allow for a better understanding of the cause of UAV accidents associated with human error through recreation of accident scenarios and evaluation of UAV pilot commands. This setup stems from discussions with cognitive psychologists on a phenomenon called shared fate. The hypothesis explains that because the ground operator does not share the same fate as the UAV flying in the air, the operator often makes overly aggressive maneuvers that increase the likelihood of crashes. During the experiments, motion cues will be given to the pilot inside the cockpit of the motion platform based on the angular rates of the UAV. The current goals of the experiments will be to assess the following questions in regards to motion cueing:

1. What skills during UAV tasks are improved/degraded under various conditions?
2. To what degree does prior manned aircraft experience improve/degrade control of the UAV?
3. How does it affect a UAV pilot's decision making process and risk taking behaviors due to shared fate sensation?

This paper is part one of a three part development of a novel UAV flight training setup that allows for pilot evaluation and can seamlessly transition pilots into a

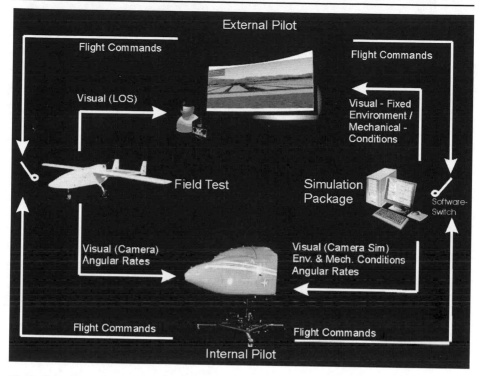

Fig. 2 Experimental setup for evaluating the effectiveness of motion cueing for UAV control. The benefit of this system is that pilots learn on the same system for simulation as they would use in the field

mission capable system. Part two will be the research to assess the effectiveness of the system and Part three will be the presentation of the complete trainer to mission ready system. As such, this paper presents the foundation of the UAV system which includes the software interface for training and the hardware interface for the mission capable system. Figure 2 shows the system and its general parts. This paper explores the use of motion platforms that give the UAV pilot increased awareness of the aircraft's state. The middle sections motivate this paper further by presenting studies on UAV accidents and how these aircraft are currently flown. It details the setup for simulation, training, human factor studies and accident assessment and presents the tele-operation setup for the real-world field tests. The final sections present and discuss experimental results, the conclusions and outlines future work.

2 UAV Operation and Accidents

While equipment failure has caused some of the accidents, human error has been found to be a significant causal factor in UAV mishaps and accidents [6, 7]. According to the Department of Defense, 70% of manned aircraft non-combat losses are attributed to human error, and a large percentage of the remaining losses have human error as a contributing factor [6]. Many believe the answer to this problem is full autonomy. However, with automation, it is difficult to anticipate all possible contingencies that can occur and to predict the response of the vehicle to all possible events. A more immediate impact can be made by modifying the way that a pilot is trained and how they currently control UAVs [8].

Many UAV accidents occur because of poor operator control. The current modes of operation for UAVs are: (1) external piloting (EP) which controls the vehicle by line of sight, similar to RC piloting; (2) internal piloting (IP) using a ground station and on board camera; and (3) autonomous flight. Some UAV systems are operated using a single mode, like the fully autonomous Global Hawk. Others are switched between modes like the Pioneer and Mako. The Pioneer used an EP for takeoff/landing and an IP during flight from a ground station. The current state of the art ground stations, like those for the Predator, contain static pilot and payload operator consoles. The pilot controls the aircraft with a joystick, rudder pedals and monitoring screens, one of which displays the view from the aircraft's nose.

The internal pilot is affected by many factors that degrade their performance such as limited field of view, delayed control response and feedback, and a lack of sensory cues from the aircraft [7]. These factors lead to a low situational awareness and decreased understanding of the state of the vehicle during operation. In turn this increases the chance of mishaps or accidents. Automating the flight tasks can have its draw backs as well. In a fully autonomous aircraft like the Global Hawk, [9] showed that because of the high levels of automation involved, operators do not closely monitor the automated mission-planning software. This results in both lowered levels of situational awareness and ability to deal with system faults when they occurred.

Human factors research has been conducted on UAV ground station piloting consoles leading to proposals on ways to improve pilot situational awareness. Improvements include new designs for head up displays [10], adding tactile and haptic feedback to the control stick [11, 12] and larger video displays [13]. To the author's knowledge, no research has been conducted in the use of motion cueing for control in UAV applications.

Potential applications of civilian UAVs such as search and rescue, fire suppression, law enforcement and many industrial applications, will take place in near-Earth environments. These are low altitude flying areas that are usually cluttered with obstacles. These new applications will result in an increased potential for mishaps. Current efforts to reduce this risk have been mostly focused on improving the autonomy of unmanned systems and thereby reducing human operator involvement. However, the state of the art of UAV avionics with sensor suites for obstacle avoidance and path planning is still not advanced enough for full autonomy in near-Earth environments like forests and urban landscapes. While the authors have shown that UAVs are capable of flying in near-Earth environments [14, 15], they also emphasized that autonomy is still an open challenge. This led the authors to focus less on developing autonomy and more on improving UAV operator control.

3 Simulation and Human Factor Studies

There are a few commercial UAV simulators available and the numbers continue to grow as the use of UAV's becomes more popular. Most of these simulators are developed to replicate the state of the art training and operation procedures for current military type UAVs. The simulation portion of our system is designed to train pilots to operate UAVs in dynamic environment conditions utilizing the motion feedback we provide them. The simulation setup also allows for reconstruction of UAV accident scenarios, to study in more detail of why the accident occurred, and allows for the placement of pilots back into the accident situation to train them on how to recover. The simulation utilizes the same motion platform and cockpit that would be used for the real world UAV flights so the transfer of the training skills to real world operation should be very close to 100%.

3.1 X-Plane and UAV Model

The training system utilizes the commercial flight simulator software known as X-Plane from Laminar Research. Using commercial software allows for much faster development time as many of the necessary items for simulation are already packaged in the software. X-Plane incorporates very accurate aerodynamic models into the program and allows for real time data to be sent into and out of the program. X-Plane has been used in the UAV research community as a visualization and validation tool for autonomous flight controllers [16]. In [16] they give a very detailed explanation of the inner workings of X-Plane and detail the data exchange through UDP. We are able to control almost every aspect of the program via two methods. The first method is an external interface running outside of the program created in a Visual Basic environment. The external program communicates with X-Plane through UDP. The second method is through the use of plug-ins developed using the X-Plane software development kit (SDK) Release 1.0.2 (freely available from http://www.xsquawkbox.net/xpsdk/). The X-Plane simulator was modified to fit this project's needs. Through the use of the author created plug ins, the simulator is capable of starting the UAV aircraft in any location, in any state, and under any condition for both an external pilot and an internal pilot. The plugin interface is shown on the right in Fig. 5. The benefit of the plugin is that the user can start the aircraft in any position and state in the environment which becomes beneficial when training landing, accident recovery and other in air skills. Another added benefit of the created plugin is that the user can also simulate a catapult launch by changing the position, orientation, and starting velocity of the vehicle. A few of the smaller UAVs are migrating toward catapult launches [17]. Utilizing X-Plane's modeling software, a UAV model was created that represents a real world UAV currently in military operation. The Mako as seen in Fig. 3 is a military drone developed by Navmar Applied Sciences Corporation. It is 130 lb, has a wingspan of 12.8 ft and is operated via an external pilot for takeoff and landings. The vehicle is under computer assisted autopilot during flight. For initial testing, this UAV platform was ideal as it could be validated by veteran Mako pilots in the author's local area. Other models of UAVs are currently available online such as the Predator A shown on the right in Fig. 3. The authors currently have a civilian Predator A pilot evaluating the accuracy of the model. The trainer is setup for the Mako such that an external pilot can train on flight tasks using an external view and RC control as in normal operation seen in Fig. 4. The system is then capable of switching to an internal view (simulated nose camera as seen in Fig. 4) at any moment to give control and send motion cues to a pilot inside of the motion platform.

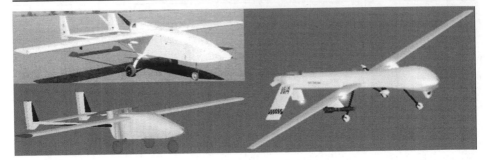

Fig. 3 *Top left* Mako UAV developed by NAVMAR Applied Sciences. *Bottom left* Mako UAV recreated in X-Plane. *Right* predator A model created by X-Plane online community

3.2 Human Factor Studies

Discussions with experienced UAV pilots of Mako and Predator A & B UAVs on current training operations and evaluation metrics for UAV pilots has helped establish a base from which to assess the effectiveness of the proposed motion integrated UAV training/control system.

The external pilot of the Mako and internal pilot of the Predator systems learn similar tasks and common flight maneuvers when training and operating the UAVs. These tasks include taking off, climbing and leveling off. While in the air, they conduct traffic pattern maneuvering such as a rectangular course and flight maneuvers such as Dutch rolls. On descent, they can conduct traffic pattern entry, go around procedures and landing approaches. These tasks are conducted during training and mission operations in various weather, day and night conditions. Each condition requires a different skill set and control technique. More advanced training includes control of the UAV during different types of system failure such as engine cutoff or camera malfunction. Spatial disorientation in UAVs as studied by [18] can effect both internal and external pilots causing mishaps. The simulator should be able to train pilots to experience and learn how to handle spatial disorientation without the financial risk of losing an aircraft to an accident.

Fig. 4 Simulator screen shots using the Mako UAV model. *Left* external pilot view point with telemetry data presented on screen. In the real world, this data is normally relayed to the pilot through a headset. *Right* internal view point with telemetry data presented. The view simulates a nose camera position on the aircraft and replicates the restricted field of view

Assessing the effectiveness of integrating motion cueing during piloting of a UAV will be conducted by having the motion platform provide cues for yaw, pitch and roll rates to the pilots during training tasks listed earlier. During simulation, the motion cues will be based on aircraft state information being fed out of the X-Plane simulation program. During field tests, the motion cues will be received wirelessly from the inertial measurement unit (IMU) onboard the aircraft. The proposed subjects will be groups of UAV internal pilots (Predator) with manned aircraft experience, UAV internal pilots without manned aircraft experience, and UAV external pilots without manned aircraft experience.

Results from these experiments will be based on quantitative analysis of the recorded flight paths and control inputs from the pilots. There will also be a survey given to assess pilot opinions of the motion integrated UAV training/control system. The work done by [19] offers a comprehensive study addressing the effects of conflicting motion cues during control of remotely piloted vehicles. The conflicting cues produced by a motion platform were representative of the motion felt by the pilot when operating a UAV from a moving position such as on a boat or another aircraft. Rather than conflicting cues, the authors of this paper will be studying the effects of relaying actual UAV motion to a pilot. We are also, in parallel, developing the hardware as mentioned earlier for field testing to validate the simulation. The authors feel that [19] is a good reference to follow for conducting the human factor tests for this study.

3.3 X-Plane and Motion Platform Interface

The left side of Fig. 5 shows the graphical user interface (GUI) designed by the authors to handle the communication between X-Plane and the motion platform ground station described in a later sections. The interface was created using Visual Basic 6 and communicates with X-Plane via UDP. The simulation interface was designed such that it sends/receives the same formatted data packet (via 802.11) to/from the motion platform ground station as an IMU would during real world flights. This allows for the same ground station to be used during simulation and field tests without any modifications. A button is programmed into the interface that allows either the attached RC controller command of the simulated UAV or the pilot inside the motion platform command at any desired moment. This would represent the external pilot control of the vehicle (RC controller) and the internal pilot control (from inside the motion platform) that would be typical of a mission setup. Currently the authors are sending angular rate data from X-Plane to the motion platform ground station and reading back into X-Plane the stick commands from the internal pilot inside the motion platform cockpit. Another powerful aspect of the program interface is that it allows the user to manipulate the data being sent out of and back into X-Plane. Noise can be easily added to the data, replicating real-world transmissions from the IMU. Time lag can also be added to data going into and out of X-plane which would represent real world data transmission delay. For example, Predator and other UAV pilots have seen delays on the order of seconds due to the long range operation of the vehicle and the use of satellite communication links [20]. Inexperienced pilots of the Predator have experienced pilot induced oscillations due to the time lag which has been the cause of some UAV mishaps.

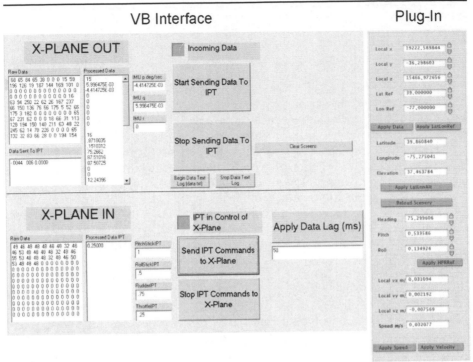

Fig. 5 *Left* graphical user interface for communication between X-Plane and IPT ground station. *Right* plugin interface running inside of X-Plane

4 Tele-operation Setup

The tele-operated system is made up of five major parts: (1) the motion platform, (2) the aerial platform, (3) the on board sensors including wireless communication, (4) the PC to remote control (RC) circuit and (5) the ground station.

4.1 Motion Platform

To relay the motion of the aircraft to the pilot during both simulation and field tests, the authors utilized a commercially available 4-*dof* flight simulator platform from Environmental Tectonics Corporation (ETC) shown in Fig. 6. ETC designs and manufactures a wide range of full-motion flight simulators for tactical fighters, general fixed-wing aircraft and helicopters. For initial development, a 4-*dof* integrated physiological trainer (IPT) system was employed because of its large workspace and fast accelerations. These are needed to replicate aircraft flight. The motion system capabilities are shown in Table 1. The cockpit is modified for specific aircrafts offering a high fidelity experience to the pilot. The visual display inside the motion platform can handle up to a 120° field of view. Basic output from the motion platform utilized in this work are the flight commands from the pilot in the form of encoder positions of the flight stick (pitch and roll), rudder pedals (yaw), and throttle.

Fig. 6 IPT 4-*dof* motion
platform from ETC being
wirelessly controlled with
the MNAV

The motion platform generates the appropriate motion cues to the pilot based on the angular velocities that it receives from the ground station. Motion cues are brief movements in the direction of acceleration which give the sensation of constant motion to the pilot but are "washed out" before the motion platform exceeds its reachable workspace. Washout algorithms are commonly used by the motion platform community to return the platform to a neutral position at a rate below the threshold that humans can sense [21]. This allows the platform to simulate motions much greater than its reachable workspace. For the IPT motion platform in particular, angular rate data streaming from the MNAV is filtered and then pitch and roll rates are washed out. The yaw rate is fed straight through due to the continuous yaw capabilities of the IPT motion platform.

4.2 Aerial Platform

The authors are particularly interested in UAV rotorcraft because they are well suited to fulfill missions like medevac and cargo transport which demand hovering, pirouettes and precision positioning. For proof of concept, the immediate goal was

Table 1 Select ETC GYRO IPT II motion system capabilities

Degree of freedom	Displacement	Speed	Acceleration
Pitch	±25°	0.5–25°/s	0.5–50°/s^2
Roll	±25°	0.5–25°/s	0.5–50°/s^2
Continuous yaw	±360° continuous	0.5–150°/s	0.5–15°/s^2

For complete specs please see ETC website

Fig. 7 The Sig Giant Kadet
model aircraft used as the
testing platform

to ensure a master-slave setup where the UAV's motions can be reproduced (in real-time) on a motion platform. To build system components, a fixed-wing UAV was used for initial demonstrations.

Rather than start with a Mako which costs on the order of thousands of dollars, the Sig Kadet offers a much cheaper, and quicker crash recovery solution for initial tests. With the Sig Kadet, the proper sensor suite and communication issues can be worked out before switching to an aircraft like the Mako shown in the earlier simulation section of this paper. The Sig Kadet shown in Fig. 7 is a very stable flight platform and is capable of carrying a sensor suite and camera system. It uses five servo motors controlled by pulse position modulated (PPM) signals to actuate the elevator, ailerons, rudder and throttle. With its 80 in. wingspan, it is comparable in size to the smaller back packable UAVs like the FQM-151 Pointer and the Raven [17].

4.3 On Board Sensors

On board the aircraft is a robotic vehicle sensor suite developed by Crossbow inertial systems. The MNAV100CA (MNAV) is a 6-*df* inertial measurement unit (IMU) measuring on board accelerations and angular rates at 50 Hz. It is also capable of measuring altitude, airspeed, GPS and heading. The MNAV is attached to the Stargate, also from Crossbow, which is an on board Linux single board computer. The Stargate is set to transmit the MNAV data at 20 Hz to the ground station via a wireless 802.11 link. As shown in Fig. 8, the MNAV and Stargate fit inside the cockpit of the Sig Kadet close to the aircraft's center of gravity.

On board video is streamed in real time to the ground station via a 2.4 GHz wireless transmission link. The transmitter is held under the belly of the Sig Kadet and the camera is located off the left wing of the aircraft. The current camera used has a 70° field of view and is capable of transmitting images at 30 FPS and 640 × 480 to a distance of 1.5 miles (AAR03-4/450 Camera from wirelessvideocameras.net). This is relatively low quality as compared with high definition camera systems but it is inexpensive, making it a decent choice for initial tests. Future tests will include much higher resolution cameras for a better visual for the pilots and a more strategic placement of the camera to replicate a pilot's on board view.

Fig. 8 MNAV and Stargate
in the cockpit of the aircraft
(top view)

MNAV+Stargate

4.4 PC to RC

Encoder positions of the flight stick, rudder pedals, and throttle inside the motion platform are transmitted via an Ethernet link to the ground station. The signals are then routed through a PC to RC circuit that converts the integer values of the encoders to pulse position modulated (PPM) signals. The PPM signals are sent through the buddy port of a 72 MHz RC transmitter which then transmits the signal to the RC receiver on board the aircraft. The PPM signals are routed to the appropriate servos to control the position of the ailerons, elevator, rudder, and throttle of the aircraft. The positions of the IPT flight controls are currently sent through the PC to RC link at a rate of 15 Hz.

4.5 Ground Station

The ground station used for the tele-operation system is a highly modified (by the authors) version of the MNAV Autopilot Ground station freely distributed on SourceForge.net. The modified ground station does three things. (1) It receives all the information being transmitted wirelessly from the MNAV and displays it to the user operating the ground station. (2) It acts as the communication hub between the aircraft and the motion platform. It relays the MNAV information via Ethernet link to the motion platform computers and sends the flight control positions of the motion platform to the PC to RC circuit via USB. (3) It continuously monitors the state of the communication link between the motion platform and the MNAV. If something fails it will put both the motion platform and aircraft (via the MNAV/Stargate) into a safe state. Determining if the ground station responds to an IMU or X-Plane data packets is set by assigning either the IP address of the IMU or the IP address of the simulator in the IPT ground station.

4.6 Field Tests

Current field tests have been conducted at a local RC flying field with the aircraft under full RC control. The field is approximately a half mile wide and a quarter mile deep. Avionics data such as angular velocity rates, accelerations and elevation was collected and recorded by the MNAV attached to the aircraft during flight. Video from the onboard camera was streamed wirelessly to the ground station and recorded. During each flight, the RC pilot conducted take off, figure eight patterns and landing with the Sig Kadet.

5 Initial Test Results and Discussion

As of writing this paper, the simulation portion was coming to completion and preparing for pilot testing and verification. In this section, the authors will present initial test results from the hardware control portion of the UAV system. In this prototyping stage, development was divided into three specific tasks that include: (1) motion platform control using the MNAV, (2) control of the aircraft servos using the IPT flight controls and (3) recording of actual flight data from the MNAV and replay on the IPT.

5.1 Motion Platform Control with MNAV

Aircraft angular rates are measured using the MNAV and this information is transmitted down to the ground station via a 20 Hz wireless link. Task A demonstrated the MNAV's ability to communicate with the ground station and the IPT. The MNAV was held in hand and commanded pitch, roll and yaw motion to the IPT by rotating the MNAV in the pitch, roll and yaw directions as seen in Fig. 6 (showing pitch).

Motions of the MNAV and IPT were recorded. Figure 9 shows a plot comparing MNAV and IPT data. The IPT is designed to replicate actual flight motions and therefore is not capable of recreating the very high angular rates commanded with the MNAV during the hand tests in the roll and pitch axis. The IPT handles this by decreasing the value of the rates to be within its bandwidth and it also filters out some of the noise associated with the MNAV sensor. Overall, the IPT tracked the motion being commanded by the MNAV fairly well. The IPT is limited by its reachable work space which is why the amplitude of the angular rates does not match at times.

Of considerable interest is the lag between the commanded angular rates and the response from the IPT motion platform, particularly with the yaw axis. This may be a limitation of the motion platform and is currently being assessed. Minimal lag is desired as significant differences between the motion cues from the IPT and visuals from the video feed will cause a quick onset of pilot vertigo.

5.2 Control of Aircraft Servos

Transmitting wirelessly at 15 Hz, no lag was observed between the instructor's flight box commands and the servo motor response. This is significant because it means

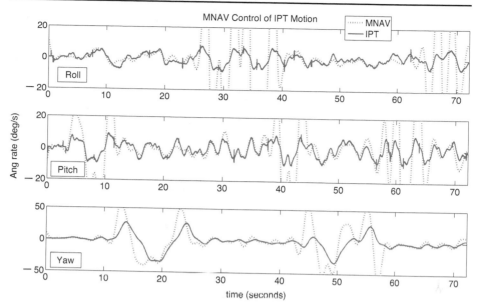

Fig. 9 Comparison of the angular rates during MNAV control of the IPT

that the pilot sitting inside the motion platform can control the aircraft through the RC link. This underscores fidelity; the aircraft will respond as if the pilot was inside its cockpit and flying the aircraft. This has only been tested during line of sight control. RC is limited in range and as stated earlier, satellite communication links for long range distances can introduce delays in data transfer. However the authors imagine near-Earth UAV applications will be conducted with groundstations near the operation site.

5.3 Record and Replay Real Flight Data

Task A demonstrated that the MNAV is able to transmit motion data to the IPT. During this task the MNAV was subjected to extreme rates and poses. Such extremes are not representative of actual aircraft angular rates but serve to demonstrate master-slave capability. To test the IPT's ability to respond to actual aircraft angular rates being sent from the MNAV, angular rate data was recorded directly from a field flight of the Sig Kadet. This data was replayed on the IPT along with on board flight video. The recorded video and flight data simulate the real time streaming information that would occur during a field tele-operation experiment. An example of the recorded angular rates from one of the field tests is shown in Fig. 10 and a still shot of the on board video recording is shown in Fig. 11.

Initial results showed errors in the angular rates between the observed motion and the recorded data. For example, the pitch rate (Fig. 10), while it is oscillating, rarely

Fig. 10 Filtered angular rates during actual aircraft flight

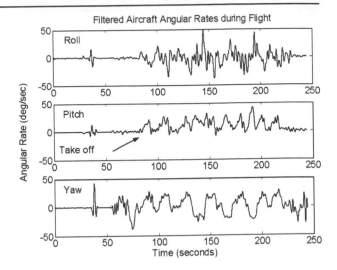

goes negative. This means that the sensor is measuring a positive pitch rate during most of the flight. Comparison of the rates with onboard aircraft video shows the error varying throughout the data so it is not a simple offset fix. This was consistently the case for multiple flights. The authors emphasize that this phenomenon was only seen during flights. Hand held motions always produced correct and expected angular rates. The recorded flight data was replayed on the IPT motion platform. This caused the IPT to travel and remain at its kinematic joint limits as was expected because of the aforementioned positive pitch rate.

The IMU was re-visited to output angular rates that reflect the bias correction made in the Kalman filter for the rate gyros [22]. A plot of the biases during a real flight is shown in Fig. 12. The resulting biases were very small and did little to fix the positive pitch rate phenomenon during flights. Alternative IMUs are thus

Fig. 11 Onboard camera view off of the left wing during flight

Fig. 12 Rate gyro biases
during actual aircraft flight

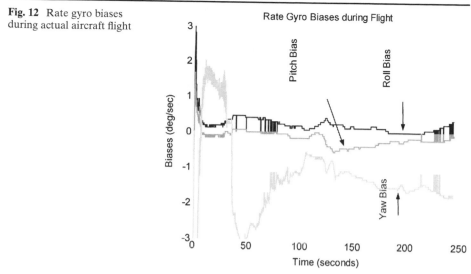

being explored at this prototyping stage. None the less, the integration of an IMU and motion platform was successfully developed. This underscores that the wireless communication interface and low-level avionics work as designed.

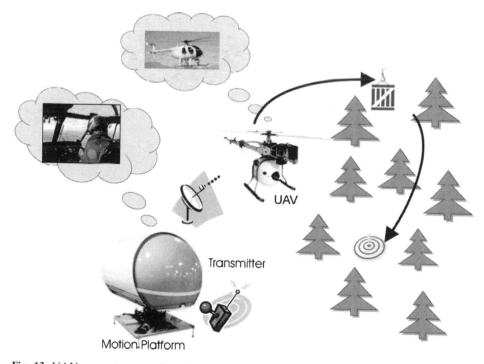

Fig. 13 UAV cargo transport in a cluttered environment using a radio link that slaves robotic helicopter motions to the motion platform. Through a "shared fate" sensation the pilot flies by "feeling" the UAV's response to maneuvers commanded by the pilot

6 Conclusion and Future Work

While the future of UAVs is promising, the lack of technical standards and fault tolerant systems are fundamental gaps preventing a vertical advance in UAV innovation, technology research, development and market growth. This paper has presented the development of the first steps toward a novel tele-operation paradigm that employs motion cueing to augment UAV operator performance and improve UAV flight training. This method has the potential to decrease the number of UAV accidents and increase the applicability of unmanned technology.

Leveraging this work, future development includes research to eliminate, reduce, or compensate for the motion lag in the motion platform. Also to be examined are additional cues like sight, touch and sound that may improve UAV control. Utilizing the system for accident reconstruction will also be assessed. The net effect is that from such understanding, one can analytically design systems to better control UAVs, train UAV pilots and help eliminate UAV accidents.

The shared fate and motion cueing will have tremendous benefit in near-Earth flying. Figure 13 depicts a notional mission involving cargo pickup and transport through a cluttered terrain to a target location. The motion platform can be used to implement a virtual "shared fate" infrastructure to command a robotic helicopter. The visuals from the helicopter's on board cameras would be transmitted to the motion platform cockpit. Added cues like audio, vibration, and motion would enable the pilot to perform precision maneuvers in cluttered environments like forests or urban structures. Future work demands the look at rotorcraft because their potential applications extend beyond the capabilities of current fixed wing UAVs. There are still a number of beneficial, life saving applications that are unachievable with current UAV methods. Among these are applications such as search and rescue and fire fighting. Even cargo transport is still very difficult to achieve autonomously in non-optimal conditions and cluttered environments. These tasks require quick, precise maneuvers and dynamic mission plans due to quickly changing environment conditions and close quarter terrain. To date these missions can only be flown by experienced, on board pilots, who still incur a great deal of risk.

Acknowledgements The authors would like to thank NAVMAR Applied Sciences for their support on the development of the UAV model and granting access to UAV pilots. The authors would also like to thank Brian DiCinti for his help with the construction of the Sig Kadet and piloting the aircraft. Acknowledgment also goes out to Canku Calargun, Caglar Unlu, and Alper Kus for their help interfacing the IPT motion platform with the MNAV. Finally the authors acknowledge Bill Mitchell, president of ETC, for his generosity in donating time on the IPT Motion platform, the supporting man power, and his overall support of this project.

References

1. Flight International: Belgians in Congo to probe fatal UAV incident. 10 October (2006)
2. Weibel, R.E., Hansman, R.J.: Safety considerations for operation of unmanned aerial vehicles in the national airspace system. Tech. Rep. ICAT-2005-1, MIT International Center for Air Transportation (2005)
3. Parrish, R.V., Houck, J.A., Martin, D.J., Jr.: Empirical comparison of a fixed-base and a moving-base simulation of a helicopter engaged in visually conducted slalom runs. NASA Tech. Rep. **D-8424**, 1–34 (1977)

4. Ricard, G.L., Parrish, R.V.: Pilot differences and motion cuing effects on simulated helicopter hover. Hum. Factors **26**(3), 249–256 (1984)
5. Wiegmann, D.A., Goh, J., O'Hare, D.: The role of situation assessment and flight experience in pilot's decisions to continue visual flight rules flight into adverse weather. Hum. Factors **44**(2), 189–197 (2001)
6. Rash, C.E., Leduc, P.A., Manning, S.D.: Human factors in U.S. military unmanned aerial vehicle accidents. Adv. Hum. Perform. Cognit. Eng. Res. **7**, 117–131 (2006)
7. Williams, K.W.: Human factors implications of unmanned aircraft accidents: flight-control problems. Adv. Hum. Perform. Cognit. Eng. Res. **7**, 105–116 (2006)
8. Schreiber, B.T., Lyon, D.R., Martin, E.L., Confer, H.A.: Impact of prior flight experience on learning predator UAV operator skills. Tech. rep., Air Force Research Laboratory Human Effectiveness Directorate Warfighter Training Research Division (2002)
9. Tvaryanas, A.P.: USAF UAV mishap epidemiology, 1997–2003. In: Human Factors of Uninhabited Aerial Vehicles First Annual Workshop Scottsdale, Az (2004)
10. Williams, K.W.: A summary of unmanned aircraft accident/incident data: human factors implications. Tech. Rep. DOT/FAA/AM-04/24, US Department of Transportation Federal Aviation Administration, Office of Aerospace Medicine (2004)
11. Calhoun, G., Draper, M.H., Ruff, H.A., Fontejon, J.V.: Utility of a tactile display for cueing faults. In: Proceedings of the Human Factors and Ergonomics Society 46th Annual Meeting, pp. 2144–2148 (2002)
12. Ruff, H.A., Draper, M.H., Poole, M., Repperger, D.: Haptic feedback as a supplemental method of altering UAV operators to the onset of turbulence. In: Proceedings of the IEA 2000/ HFES 2000 Congress, pp. 3.14–3.44 (2000)
13. Little, K.: Raytheon announces revolutionary new 'cockpit' for unmanned aircraft—an industry first. In: Raytheon Media Relations (2006)
14. Sevcik, K.W., Green, W.E., Oh, P.Y.: Exploring search-and-rescue in near-earth environments for aerial robots. In: IEEE International Conference on Advanced Intelligent Mechatronics Monterey, California, pp. 699–704 (2005)
15. Narli, V., Oh, P.Y.: Hardware-in-the-loop test rig to capture aerial robot and sensor suite performance metrics. In: IEEE International Conference on Intelligent Robots and Systems, p. 2006 (2006)
16. Ernst, D., Valavanis, K., Garcia, R., Craighead, J.: Unmanned vehicle controller design, evaluation and implementation: from matlab to printed circuit board. J. Intell. Robot. Syst. **49**, 85–108 (2007)
17. Defense, D.O.: Unmanned aircraft systems roadmap 2005–2030. Tech. rep., August (2005)
18. Self, B.P., Ercoline, W.R., Olson, W.A., Tvaryanas, A.: Spatial disorientation in unihabited aerial vehicles. In: Cook, N. (ed.) Human Factors of Remotely Operated Vehicles, vol. 7, pp. 133–146. Elsevier Ltd. (2006)
19. Reed, L.: Visual-proprioceptive cue conflicts in the control of remotely piloted vehicles. Tech. Rep. AFHRL-TR-77-57, Brooks Airforce Base, Air Force Human Resources Laboratory (1977)
20. Mouloua, M., Gilson, R., Daskarolis-Kring, E., Kring, J., Hancock, P.: Ergonomics of UAV/UCAV mission success: considerations for data link, control, and display issues. In: Human Factors and Ergonomics Soceity 45th Annual Meeting, pp. 144–148 (2001)
21. Nahon, M.A., Reid, L.D.: Simulator motion-drive algorithms: a designer's perspective. J. Guid. Control Dyn. **13**, 356–362 (1990)
22. Jang, J.S., Liccardo, D.: Automation of small UAVs using a low cost mems sensor and embedded computing platform. In: 25th Digital Avionics Systems Conference, pp. 1–9 (2006)

Networking Issues for Small Unmanned Aircraft Systems

Eric W. Frew · Timothy X. Brown

Originally published in the Journal of Intelligent and Robotic Systems, Volume 54, Nos 1–3, 21–37.
© Springer Science + Business Media B.V. 2008

Abstract This paper explores networking issues that arise as a result of the operational requirements of future applications of small unmanned aircraft systems. Small unmanned aircraft systems have the potential to create new applications and markets in civil domains, enable many disruptive technologies, and put considerable stress on air traffic control systems. The operational requirements lead to networking requirements that are mapped to three different conceptual axes that include network connectivity, data delivery, and service discovery. The location of small UAS networking requirements and limitations along these axes has implications on the networking architectures that should be deployed. The delay-tolerant mobile ad-hoc network architecture offers the best option in terms of flexibility, reliability, robustness, and performance compared to other possibilities. This network architecture also provides the opportunity to exploit controlled mobility to improve performance when the network becomes stressed or fractured.

Keywords Unmanned aircraft system · UAS · Airborne communication networks · Controlled mobility · Heterogeneous unmanned aircraft system · Mobile ad-hoc networking · Delay tolerant networking

1 Introduction

The proliferation of small unmanned aircraft systems (UAS) for military applications has led to rapid technological advancement and a large UAS-savvy workforce poised

E. W. Frew (✉)
Aerospace Engineering Sciences Department, University of Colorado,
Boulder, CO 80309, USA
e-mail: eric.frew@colorado.edu

T. X. Brown
Interdisciplinary Telecommunications Program Electrical
and Computer Engineering Department, University of Colorado, Boulder, CO 80309, USA
e-mail: timxb@colorado.edu

K. P. Valavanis et al. (eds.), *Unmanned Aircraft Systems*. DOI: 10.1007/978-1-4020-9137-7_3

to propel unmanned aircraft into new areas and markets in civil domains. Small unmanned aircraft systems have already been fielded for missions such as law enforcement [29], wildfire management [34], pollutant studies [10], polar weather monitoring [11], and hurricane observation [26]. Proposed UAS span numerous more future civilian, commercial, and scientific applications. A recent study concluded that in 2017 the civil UAS market in the USA could reach $560 M out of a total (civil plus military) UAS market of approximately $5.0 B [32]. That study projects 1,500 civil UAS will be in service in 2017 and that approximately 85% of those will be small UAS.

As the number of fielded small UAS grows, networked communication will become an increasingly vital issue for small UAS development. The largest current barrier to the use of unmanned aircraft in the National Airspace System (NAS) of the USA is satisfaction of Federal Aviation Administration (FAA) regulations regarding safe flight operations and Air Traffic Control (ATC). In particular, the FAA requires almost all aircraft operating in the NAS to have a detect, sense, and avoid (DSA) capability [3] that provides an equivalent level of safety compared to manned aircraft [1, 33]. While onboard sensors are expected to be a component of future DSA solutions, communication to ATC and operator intervention will also be required, either from a regulatory or practical perspective. Thus, one of the primary concerns of the FAA regarding the ability of UAS to meet safety regulations without conflicting with existing systems is the availability and allocation of bandwidth and spectrum for communication, command, and control [2]. Although the particular regulations just mentioned refer to operation in the USA, similar concerns apply to the operation of small UAS anywhere.

Small unmanned aircraft (UA) are defined here to encompass the Micro, Mini, and Close Range categories defined in [5]. This classification means small UA have maximum takeoff weight less than or equal to 150 kg, maximum range of 30 km, and maximum altitude of 4,000 m mean sea level (MSL). The weight limit effectively means small UA can not carry the equivalent weight of a human operator. The altitude limit taken here means small UA cannot fly into Class A airspace (the airspace from 18,000 to 60,000 ft MSL where commercial aircraft fly). Although it may be possible for vehicles in this category to fly at higher altitudes, the regulatory issues are significantly more challenging and it is reasonable to assume most small UA will not fly in that airspace. In fact, most small UA would probably fly substantially closer to the ground. Likewise, the maximum range of 30 km represents typical operational limits on this class of aircraft and there can be notable exceptions [11]. Finally, note that a small UAS can be comprised of multiple heterogeneous small UA with highly varying capabilities.

Unlike larger unmanned aircraft, small UAS are in a unique regime where the ability to carry mitigating technology onboard is limited yet the potential for damage is high. Given the size and payload constraints of small UAS, these unmanned aircraft have limited onboard power, sensing, communication, and computation. Although the payload capacity of a small UAS is limiting, the kinetic energy stored in a 150 kg aircraft can cause significant damage to other aircraft, buildings, and people on the ground. Furthermore, the limited sizes of small UAS make them accessible to a wider audience (e.g. a variety of universities already have small UAS programs [4, 9, 16, 21, 28]) than larger systems and the percentage of small UAS deployed in the future will likely be high relative to larger unmanned aircraft systems [32]. The limited capabilities of small UAS lead to unique operational requirements compared to larger UA that can more easily assimilate into the existing ATC framework (e.g. larger UA can carry the same transponder equipment as manned aircraft).

This paper explores networking issues that arise as a result of the operational requirements of future applications of small unmanned aircraft systems. These requirements are derived from a set of representative application scenarios. The operational requirements then lead to networking requirements (e.g. throughput, which is the rate at which data can be sent over a communication link, and latency or delay) that greatly exceed those of current manned aircraft. Further, the networking requirements are mapped to three different conceptual axes that include network connectivity, data delivery, and service discovery. The location of small UAS networking requirements and limitations along these axes has implications on the networking architectures that should be deployed.

Of the existing possible network architectures for small UAS, only delay-tolerant mobile ad-hoc networking architectures will provide the needed communication for the large number of small aircraft expected to be deployed in the future. Since small UA are relatively cheap, future UAS will likely deploy multiple vehicles coordinated together. Many small UAS applications will require quick response times in areas where permanent supporting communication infrastructures will not exist. Furthermore, current approaches using powerful long-range or satellite communications are too big and expensive for small aircraft while smaller radios fundamentally limit the small UAS operational envelope in terms of range, altitude, and payload. The delay-tolerant mobile ad-hoc network architecture offers the best option in terms of flexibility, reliability, robustness, and performance compared to other possibilities. This network architecture also provides the opportunity to exploit controlled mobility to improve performance when the network becomes stressed or fractured.

2 Communication Requirements

2.1 Operational Requirements

This work is motivated by the Heterogeneous Unmanned Aircraft System (HUAS) developed at the University of Colorado as a platform to study airborne communication networks and multivehicle cooperative control (Fig. 1 shows the various small UA included in HUAS). Specific applications studied to date include the impact of mobility on airborne wireless communication using off the shelf IEEE 802.11b (WiFi) radios [7]; net-centric communication, command, and control of small UAS [16]; sensor data collection [22]; delay tolerant networking [6]; and a framework for controlled mobility that integrates direct, relay, and ferrying communication concepts [13].

As an example application consider a UAS to track a toxic plume. In this scenario a toxic plume has been released in an accident and the goal is to locate the plume extent and source [17]. To characterize the plume, multiple small UA fly while sensing the plume with onboard chemical sensors. Different sensors may be in different UA because it may not be possible or desirable for every small UA to carry every sensor. UA with the same chemical sensors onboard need to find each other to form gradient seeking pairs. Chemical gradients can be defined by sharing sensor

Fig. 1 The HUAS vehicle fleet includes (*clockwise from top left*) the CU Ares, the CU MUA, the Velocity XL, the MLB Bat 3, the CU ground control station, and the Hobico NextGen

data and the UAS can cooperatively track boundaries or follow the gradients to the source. UA can potentially move far away from their launching ground stations and each other. In this example, the UAS consists of potentially many heterogeneous UA. They need to fly freely over a large area and be able to dynamically form associations autonomously without relying on a centralized controller.

As a second example consider a UAS deployed to act as a communication network over a disaster area. Here, normal communication infrastructure has been damaged but various entities on the ground such as first responders, relief agencies, and local inhabitants require communication in order to organize a coordinated response. An unmanned aircraft system flying overhead can provide a meshed communication architecture that connects local devices, e.g. laptops with wireless networking or cell phones, with each other or back to the larger communication grid. Since communication demand will vary as the severity of the disaster is assessed and relief efforts are mounted, the UAS must be able to reposition itself in response. Since the actions of the ground units are in direct response to the emergency situation, the actions of the UAS must be dependent on them and not limit their efforts. Also, the UAS will likely operate in the vicinity of buildings and other manned aircraft so obstacle and collision avoidance will be critical.

The two scenarios described above share properties with many other potential small UAS applications and lead to communication requirements for the UAS itself. In particular, these communication needs can be broadly classified into platform safety, remote piloting, and payload management (Fig. 2). In general, the UAS will communicate with multiple external parties that could include ATC, the pilot, and payload operators who may be in widely separate locations.

2.1.1 Platform Safety

From a regulatory perspective, platform safety is the most critical component of an unmanned aircraft system. Like a manned aircraft, the pilot of the UAS must communicate with ATC in most controlled airspace [33]. This communication may be mediated by the UAS whereby the UAS communicates via conventional radio

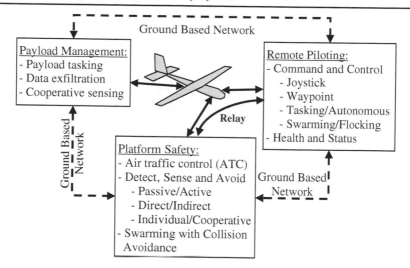

Fig. 2 Types of communication in an unmanned aircraft system

to the ATC and this communication is then backhauled to the pilot. This implies an inherent inefficiency. A single ATC radio channel is shared by all planes in an area. But, each UA requires a separate backhaul to its respective operator, and multiplies the communication requirements. Future non-voice approaches to managing aircraft are being contemplated [14]. In principle, ATC commands (e.g. to change altitude) could be acted upon directly by the UA without pilot intervention obviating the need for the inefficient backhaul. However, it is likely that a pilot will always be expected to be "on the loop" so that UAS operations are completely transparent to ATC. Future communication system analysis estimates the average ATC voice and data rates to be about 10 kbps per aircraft particularly in autonomous operations areas typical of UAS operations [14].

Other platform safety communication is related to detect, sense, and avoid requirements which generally require the UAS to have equivalent ability to avoid collisions as manned aircraft [1, 33]. This may require onboard radar (active sensing), backhaul of image data (passive sensing), transponders, or cooperative sharing of information between UA. The communication requirements here can depend significantly on the approach. The least communication demands are required when the aircraft uses active sensing, only reports potential collisions to the operator, and autonomously performs evasive maneuvers when collisions are imminent. The communication requirements here are negligible. More demanding systems send full visual situational awareness to the operator which can require 1 Mbps or more (e.g. Fig. 3).

The communication requirements for small UAS are simplified in many instances. Small UAS often fly in uncontrolled airspace. No transponders are required, nor is communication with ATC. Small UAS can operate directly in uncontrolled airspace without the need of an airport. They are often catapult or hand launched and can have parachute, net, or snag recovery systems. Small UAS generally fly over smaller regions than larger UAS. Small UAS are still subject to DSA requirements, however, for very short ranges, the pilot or spotters on the ground can provide the see and

Fig. 3 Situational awareness provide by imagery from onboard a small UA

avoid. For larger ranges active or passive techniques are required. The smaller platform size limits the ability to carry onboard autonomous DSA systems.

Small UAS participating in the example scenarios described in Section 2.1 will clearly be operating in environments with significant other air traffic so platform safety will be important. Manned aircraft for emergency response and from news agencies will surely operate in the environment where the UAS will be deployed. The UAS pilot will require significant communication with ATC to coordinate operation with these other aircraft. From the perspective of network or radio bandwidth and throughput, the requirements for this communication traffic are low since messages are limited in size (or length) and are sent sporadically. However, the safety critical nature of this traffic will require high reliability with low latency.

2.1.2 Remote Piloting

Remote piloting of the vehicle has requirements that vary with the type of flight control. On one extreme is direct joystick control of the aircraft. This requires low delay and high availability. At the other extreme, tasking commands are sent to the aircraft which are autonomously translated to flight paths (Fig. 4 shows the user interface for the Piccolo autopilot that allows for point and click commands [31]). Here delays can be longer and gaps in availability can be tolerated. The UA to

Fig. 4 Cloud cap technologies piccolo autopilot command center [31]

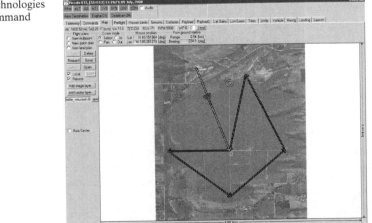

pilot link contains not only commands from the pilot to the UA but also essential health and status information from the aircraft back to the pilot. As examples, on the joystick end of the spectrum commercial digital radio control (R/C) links have data rates below 10 kbps and one-way delays below 100 ms are preferred. On the autonomous end of the spectrum, an Iridium satellite link is sufficient for waypoint flying of the Predator UAS. Iridium has 2.1 kbps throughput, delays of 1–7 s, and has gaps in connectivity with 96% average availability [25].

Small UAS are lower cost and are more likely to operate in cooperative groups. There is a strong push to enable one-to-many pilot-aircraft interactions for UAS [27]. This mode of operation would require increased amounts of UA autonomy with the pilot in charge of higher level mission planning and tasking. As such, UA must be capable of autonomous collision avoidance and therefore plane-to-plane communication becomes another significant communication component. Collision avoidance between two UA will also have low data rate and low latency requirements. However, the presence of multiple vehicles all performing plane-to-plane communication complicates the networking and introduces the need for bandwidth and congestion control. The possibility of varied capabilities and aircraft attrition also necessitates dynamic service discovery routines whereby system capabilities can be updated internally.

2.1.3 Payload Management

Communication with the payload can range from a few bits per second for simple sensor readings to megabits per second for high-quality images (Fig. 5). For instance, the Predator uses a 4.5 Mbps microwave link to communicate payload imagery when in line-of-site of the ground station [23]. The types of payload communication needed by small UAS can be highly varied. For example, depending on the type of chemical plume being tracked, real-time data assimilation may not be needed. In that case large amounts of data can be stored at intermediate nodes and transmitted opportunistically back to the end user. In contrast, if a toxic substance is released in an urban setting, source localization could take priority over all other requirements including DSA. Using multiple UA to provide information to multiple dispersed users also necessitates dynamic service discovery routines.

In summary, the communication requirements for UAS are modest for ATC communication and remote piloting while UAS can potentially require data rates in the megabits per second for payload management and DSA. It is this requirement for multiple connections, some of which are at high data rates, that distinguishes UAS from manned aircraft communications. There are also other considerations

Fig. 5 High quality payload imagery from the MLB Bat small UAS [24]

than data rates, latency, and availability. ATC, remote piloting, and other flight safety communication will likely be required to operate in protected spectrum that is not shared with payload and non-essential communication [14, 20].

2.2 Operational Networking Requirements

The communication needs can be explored along three axis (Fig. 6). The first is connectivity. In traditional networks, node connectivity is well defined; a physical wire connects two nodes. These links are designed to be reliable with rare transmission errors. Further, the links and nodes are stable with only rare failures. This yields a well defined network topology with clear notions of graph connectivity that is consistent across network nodes. In a small UAS, the links are less reliable wireless links with connectivity that ranges from good when nodes are close to poor for nodes that are further away. Even when connectivity is good packet error rates are high relative to wired standards. The transmission is broadcast and can reach multiple receivers so that connections are not simple graph edges. Further, broadcast transmissions interfere with each other so that the ability of two nodes to communicate depends on the other transmissions at the same time. As we add UA mobility, connectivity becomes dynamic and as UA speeds increase relative to the nominal communication range different UA may have inconsistent notions of connectivity.

The second axis is data delivery. In traditional networks, connectivity is well defined and stable so that data delivery is based on an end-to-end model. For instance, with TCP protocols the source and destination end points manage data delivery over a presumed reliable and low latency network. As these connectivity assumptions break down this model of delivery is not possible. As already noted, small UAS connectivity is unreliable and dynamic. Furthermore, small UAS may become spread out over a mission so that end-to-end connectivity simply does not exist for data delivery.

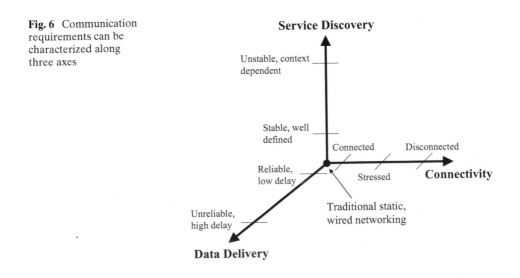

Fig. 6 Communication requirements can be characterized along three axes

The third axis is service discovery. In traditional networks, resources are stable and can be found through well defined procedures. To find a webpage, we use a URL (e.g. http://www.springer.com), this is translated by the DNS service into a network address, and we request the webpage from the server at that address. In contrast, with small UAS nodes, services, and users can come and go over the life of a mission and the resources sought may be ephemeral or context dependent. Each UA may have different capabilities (e.g. the chemical sensors in the plume tracking example) and onboard capabilities may be available for use by other UA (e.g. coordinating cameras for stereo imaging). As small UAS spread out these resources, even when they exist, may be difficult to discover. This concept of service discovery can be pushed down from aircraft to the subsystems aboard the aircraft. By connecting onboard subsystems via local subnets to the external meshed network, dispersed operators as well as other aircraft in the UAS can discover and communicate directly to different components of the aircraft avionics system.

3 Networking for Small UAS

The goal of small UAS networking is to address the communication needs given that small UAS differ from traditional networks along the connectivity, data delivery, and service discovery axes. To this end, this section describes the merits of different communication architectures and introduces the delay tolerant networking delivery mechanism. It also discusses how the mobility of the small UAS can be exploited to improve networking.

3.1 Communication Architectures

There are four basic communication architectures which can be used for small UAS applications: direct link, satellite, cellular, or mesh networking (Fig. 7). Each has advantages and disadvantages which we outline here. A direct link between the ground control station and each UA is the simplest architecture. It assumes connectivity is maintained over dedicated links to each UA and therefore data delivery is reliable with low latency. Since the ground station communicates to each UA, service discovery is easily managed by a centralized agent at the ground station. Unfortunately the direct architecture is not suited for dynamic environments and non-line-of-sight (NLOS) communication. Obstructions can block the signal, and at longer ranges the UA requires a high-power transmitter, a steerable antenna, or significant bandwidth in order to support high data rate downlinks. The amount of bandwidth scales with the number of UA so that many UAS may not operate simultaneously in the same area. Finally, plane-to-plane communication will be inefficiently routed through the ground control station in a star topology and not exploit direct communication between cooperative UA operating in the same area.

Satellite provides better coverage than a direct link to the ground control station. As a result, the UAS network can remain well connected, however this connectivity would still be provided by routing data through a centralized system. Data delivery is relatively poor using satellite. Lack of satellite bandwidth already limits existing UAS operations and will not scale with the increasing demand of 1,000s of small UAS operations in a region. For high data rate applications, a bulky steerable dish antenna mechanism unsuitable in size, weight and cost for small UAS is necessary.

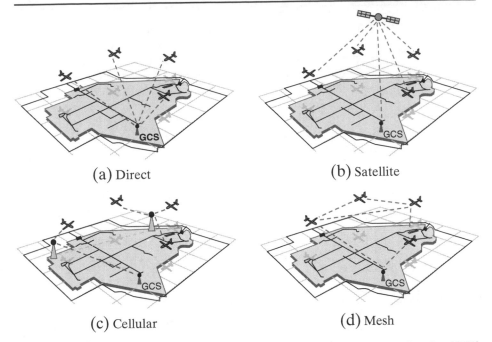

(a) Direct   (b) Satellite

(c) Cellular   (d) Mesh

Fig. 7 Four basic communication architectures for small UAS. The ground control station (GCS) represents the operator or end user

Further, the ground control station requires a connection to the satellite downlink network. The ground control station may have obstructed satellite views because of terrain or clutter. Finally, multiple UA operating in an area will suffer high delays if their communication is mediated by satellite.

Cellular refers to an infrastructure of downlink towers similar to the ubiquitous mobile telephone infrastructure. The cellular architecture has several advantages that can provide good levels of network connectivity and reliable data delivery. First, coverage can be extended over large areas via multiple base stations. UAS would hand-off between different base stations as needed during flight. Second, the multiple base stations provide a natural redundancy so that if one link is poor another link may perform better. Third, a limited bandwidth can be reused many times over a region and capacity increased as needed to meet demand. The reuse can grow by adding more base stations as the number of users grows. Fourth, the infrastructure can be shared by different UAS. Once installed, many UAS can each pay for the fraction of the infrastructure that they use. These advantages must be weighed against the cost. A typical mobile telephone base station is expensive for the tower, tower site, radio equipment, and associated networking infrastructure. Such a solution applies where the infrastructure investment can be amortized across frequent and regular UAS flights. Examples might include agricultural monitoring or border surveillance. Such architecture is not suited for applications like wildfire management or polar climatology where demand is transient. The existing mobile telephone infrastructure is not designed for air to ground communication. A single UA transmitter can blanket a large area with its signal degrading system performance. Therefore, small UAS operations may require a dedicated cellular infrastructure.

Meshing refers to a networking architecture where each node (i.e. a radio on a UA or ground node) can act as a relay to forward data. Communication between a UA and a ground control station can take place over several ŞhopsŤ through intermediate nodes. The shorter range simplifies the link requirements and bandwidth can be reused more frequently and thus more efficiently. Plane-to-plane communication can be direct and also benefit from the mesh routing protocols that employ additional relays as needed to maintain communication. However, such meshing requires intermediate nodes to be present for such relaying to take place. Furthermore, nodes may be required to move specifically in order to support communication.

Mobile ad-hoc networking (MANET) is a particular example of a mesh architecture comprised of a self-configuring network of mobile routers which are free to move randomly throughout the environment. The MANET topology changes rapidly due to the motion and the nodes in the network rapidly self-organize in response. The MANET approach is promising for UAS applications where infrastructure is not available and multiple UA are operating cooperatively. Data relaying in MANETs reduces the connectivity requirements since source and destination nodes only need to be connected through the intermediate nodes. Due to the decrease of radio transmission power, and hence communication capacity, with separation distance [30], the presence of the intermediate relay nodes can actually improve data delivery performance over direct communication [18]. Further, since MANETs are design to self-heal, they can respond well to the dynamic network topologies that result from UA motion. Service discovery in MANETs is more important and more complex than the other architectures since the network tends to become fractured for periods of time and there is no centralized node to coordinate network activities.

Meshing can also leverage the other technologies described above. A direct, satellite, or cellular link to any node in a connected mesh enables communication with all the nodes providing additional redundancy in the communication. Meshing combined with mobility can extend range. For instance, as a group of UA move beyond the range of a direct link, some of the UA can be assigned to stay behind forming a chain of links back to the direct link. In an extreme form of this UA can fly back and forth to ferry data between nodes that have become widely separated. It is this flexibility, robustness, and added range that makes meshing an integral part of any small UAS operations.

3.2 Delay Tolerant Networking

Real-time communication is challenging for mobile ad-hoc networks on small UAS because of inherent variability in wireless connections combined with the fact that nodes may become spread out and sparsely connected. Connections that exist can be dynamic and intermittent, antenna patterns shift with aircraft maneuvering, sources of interference come and go, and low flying UAS can be separated by intervening terrain. In the extreme case of sparsely connected, moving nodes, some nodes might not be able to connect with any other node for a long time. In such environments, traditional end-to-end network protocols such as the ubiquitous TCP perform poorly and only delay-tolerant communication is feasible. So-called delay tolerant networks (DTN) are designed for these challenged environments [8, 15]. DTN provide a smooth spectrum of communication ability. Data is delivered quickly when end-to-end connections exist and as quickly as opportunities appear when intermittently connected. The DTN also supports data ferrying where, for instance, a ground sensor can deliver data to an overflying UA that then physically carries the data back to a network gateway to an observer.

The data flowing through the MANETs deployed on small UAS will have large variety of data types and quality of service requirements. For example, Voice over IP (VOIP) between first responders, UA control data, or critical process monitoring will require prioritized real-time flow. For a wireless ad-hoc network to be able to carry real-time data there needs to be a contemporaneous, reliable connection from sender to receiver. The more nodes that participate in the network in a given area, the more likely a path can be established. Other data types, such as email, non-critical messaging, or sensor data from long-term experiments, will not carry the same sense of urgency and consist of delay-tolerant traffic where only eventual, reliable reception is important. Continuous multi-hop links from the sender to receiver do not need to be maintained as long as a data packet is carried or forwarded to its destination in some reasonable amount of time. The DTN architecture provides seamless mechanisms for integrating these various requirements into a single network.

DTN are a current topic of research. An example DTN was implemented on the University of Colorado's HUAS for delivering sensor data to one or more external observers outside the UAS [22]. Sensors on UA or on the ground generate data and a DTN procedure delivers the data in stages through gateways to the observers. Each stage takes custody of the data and stores it in the network until a connection opportunity to the next stage arises. End nodes can operate with no knowledge of the DTN through proxy interfaces.

In addition to extending the capability of MANET architectures to sparse or fractured networks, the DTN concept is also important for maintaining system responsibility and accountability in the face of UAS failure. In traditional MANET architectures, data is not transmitted if an end-to-end connection is not present and therefore data will be lost if a link goes down due to failure. In contrast, the DTN protocols store data at intermediate nodes until it is safely transmitted to the source. Thus telemetry, health and status, and other data collected during periods when a node is disconnected from the network can still be collected. This includes information from moments before a failure which may be stored and can be collected for post analysis.

3.3 Exploiting Controlled Mobility

Unlike terrestrial networks which tend to be static and satellite communication systems which are in fixed orbits, meshed networks of small UA offer a unique opportunity to exploit controlled mobility. Even when a continuous link is established, environmental factors such as multipath, interference, or adversarial jamming, can degrade real-time performance relative to expected models. In these cases the mobility of the nodes themselves can be exploited to improve the network performance. In sparse networks, node mobility enables data ferrying, i.e. physically carrying data packets through the environment, between otherwise disconnected nodes. In the case of a connected network, node mobility enables adjustment of local network behavior in response to unmodeled disturbances.

Given the presence of node mobility in an ad-hoc network, transmission of data between a source and destination can take three forms (Fig. 8). *Direct*

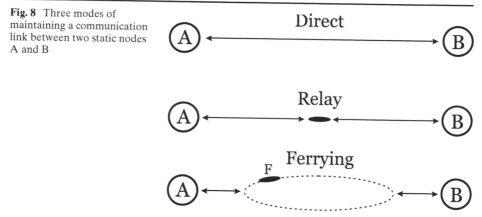

Fig. 8 Three modes of maintaining a communication link between two static nodes A and B

communication occurs when two nodes transmit data directly to one another. *Relaying* occurs when additional nodes are used to receive a transmission from a source and retransmit it to a destination. Finally, *data ferrying* occurs when a mobile node physically stores and carries data from one location to another. Each of these modes has a place in communication. For instance, at 5 km a low-power wireless direct link might support only a few 10's of kbps, while a ferry with a 50 megabyte buffer and velocity of 30 m/s can deliver data at a rate of over 1 Mbps. However, the direct link can deliver a packet in under a second while the ferry will require minutes. Thus, using mobility requires understanding the tradeoffs in delay and data rate. These ideas have been explored in [6] which defines the different operational regions and the tradeoffs in each.

To further illustrate the role of mobility in determining network performance, consider an example with two task nodes that wish to communicate with one another and additional mobile helper nodes in the environment that can act as relays between the task nodes. The two task nodes could represent a time-critical sensing system and a human-operated ground station or two relay nodes electronically leashed to convoys of ground vehicles. As the distance between the task nodes increases, one or more helper nodes are needed to relay data between them. While networking protocols controlling the flow of data between nodes have been extensively studied, the issue of how to best deploy or position these helper nodes is still open. Position-based solutions break down in the presence of noise sources and terrain that distort the radio propagation and power models from simple cases [12]. For example, Fig. 9 shows how a single noise source distorts the contours of end-to-end chain capacity as a function of relay node position for a simple 3-node network. For a multiple UA chain providing multi-hop communication, decentralized control laws can optimize chain capacity based only on measures of the local 3-node network perceived by each UA [12].

As the separation distance between the task nodes grows relative to the communication range of the helper nodes (i.e. the density scale decreases) positioning of the helper nodes becomes more difficult. Given relatively static node placement and knowledge of their position, deployment of the helper nodes becomes a resource allocation problem [18]. As the density scale decreases and the separation between nodes grows, it becomes difficult for helper nodes to build consensus on deployment.

Fig. 9 Contours of end-to-end chain capacity for a three-node network and the vector field showing the gradient of capacity. Interference distorts radio propagation from standard models and makes position-based solutions for relay placement suboptimal. **a** Contours and vector field with no noise; **b** Contours and vector field with a localized noise source

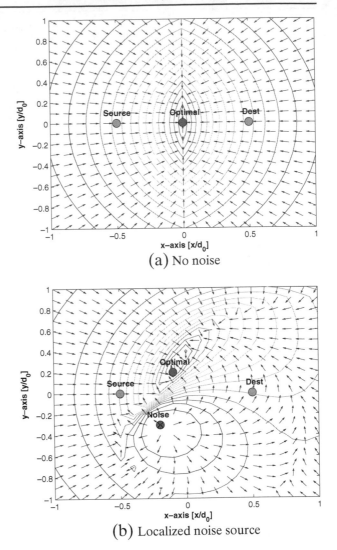

(a) No noise

(b) Localized noise source

Centralized strategies in which the helpers converge to build consensus trade advantages with distributed approaches that enable helper nodes to make decisions based only on local information. The best approach is not apparent. Furthermore, as the separation distance between nodes grows beyond a certain point it becomes impossible to establish a connected chain between them and the helper nodes must exploit their own mobility to ferry data back and forth [19]. While this seems to be a basic process, the fact that the nodes are not in a fully connected network at this point hinders consensus building among the helpers and the implementation of globally optimal behavior.

Although deployment of helper nodes is challenging in relatively static scenarios, the problem is further complicated when nodes (task nodes and helper nodes) are capable of moving quickly relative to their communication range (i.e. the dynamic scale increases). If we allow the task nodes the freedom to perform their primary task, the helper nodes must cooperatively distribute themselves through the environment to provide multi-hop communication. Motion of the helper nodes has a greater effect on the performance of the communication network and the problem shifts from resource allocation to continuous coordinated control.

This simple example scenario illustrates the effects of mobility on communication in delay tolerant networks. It is clear that the establishment and maintenance of networks of mobile nodes, especially given a subset of nodes dedicated to primary missions such as sensing, requires some form of distributed cooperative control. Furthermore, in order to exploit node mobility in a distributed control architecture, communication metrics such as quality of service requirements must be incorporated explicitly as objectives in the mobility control scheme.

4 Conclusion

This paper explored requirements for networking of small unmanned aircraft systems. The networked communication demands of unmanned aircraft systems in general, and small unmanned aircraft system in particular, are large compared to manned aircraft since telemetry, command and control, health and safety, and payload data must be sent from multiple aircraft to multiple dispersed users such as the unmanned aircraft system operator, air traffic control, and other aircraft in the vicinity.

The communication environment for small UAS was shown to deviate significantly form traditional networking assumptions. Connectivity is lower quality, less well defined, and dynamic. Data delivery can not depend on reliable, low-latency, and end-to-end connections. Network service discovery must operate in isolated networks to discover dynamically available resources.

In this environment, we described how mesh networking, supported by delay tolerant networking, and robust service discovery can operate. Mesh networking is designed to work in mobile environments and allows two UA to communicate by dynamically piecing together links to form a communication path. Delay tolerant networking is designed to work in sparse connectivity environments and inserts intermediate custody points so that communication can make progress in smaller stages over time as connections become available. Service discovery allows aircraft to advertise their capabilities and so others can find and use them.

The small UAS networking environment is challenging but provides some opportunities. A small UAS will likely have multiple aircraft. When needed some of these can be devoted to support communications, moving to locations to relay between disconnected groups or in extreme cases the aircraft can ferry data by physically carrying the data back and forth.

Small UAS provide unique challenges to networking. What is reported here is based on our experience implementing and testing networks on small UAS. Our research continues to address these challenges. These include methods for seamlessly and safely integrating flight critical command and control data with less critical payload date on a single communication network.

Acknowledgements The authors would like to thank the members of the University of Colorado AUGNet Research Group. This work was supported by the US Air Force under Grant FA9550-06-1-0205, by the Federal Aviation Administration under Grant 07-G-014, and by L-3 Communications.

References

1. Federal Aviation Administration: order 7610.4k. Special military operations (2004)
2. Federal Aviation Administration: meeting the challenge: unmanned aircraft systems. In: Federal Aviation Administration R&D Review, vol. 4 (2006)
3. Federal Aviation Administration: title 14 code of federal regulations (14 cfr) part 91 (2008)
4. Beard, R., McLain, T., Nelson, D., Kingston, D., Johanson, D.: Decentralized cooperative aerial surveillance using fixed-wing miniature UAVs. Proc. I.E.E.E. **94**(7), 1306–24 (2006)
5. van Blyenburgh, P.: Unmanned Aircraft Systems: The Global Perspective. UVS International, Paris, France (2007)
6. Brown, T.X., Henkel, D.: On controlled node mobility in delay-tolerant networks of unmanned aerial vehicles. In: Proc. of Intl Symposium on Advanced Radio Technologies, Boulder, CO (2006)
7. Brown, T.X., Argrow, B.M., Frew, E.W., Dixon, C., Henkel, D., Elston, J., Gates, H.: Experiments using small unmanned aircraft to augment a mobile ad hoc network. In: Bing, B. (ed.) Emerging Technologies in Wireless LANs: Theory, Design, and Deployment, chap. 28, pp. 123–145. Cambridge University Press (2007)
8. Cerf, V.G., Burleigh, S.C., Durst, R.C., Fall, K., Hooke, A.J., Scott, K.L., Torgerson, L., Weiss, H.S.: Delay-Tolerant Network Architecture. Internet Draft, IETF (2006)
9. Claus Christmann, H., Johnson, E.N.: Design and implementation of a self-configuring ad-hoc network for unmanned aerial systems. In: Collection of Technical Papers - 2007 AIAA InfoTech at Aerospace Conference, vol. 1, pp. 698–704. Rohnert Park, CA (2007)
10. Corrigan, C.E., Roberts, G., Ramana, M., Kim, D., Ramanathan, V.: Capturing vertical profiles of aerosols and black carbon over the Indian Ocean using autonomous unmanned aerial vehicles. Atmos. Chem. Phys. Discuss **7**, 11,429–11,463 (2007)
11. Curry, J.A., Maslanik, J., Holland, G., Pinto, J.: Applications of aerosondes in the arctic. Bull. Am. Meteorol. Soc. **85**(12), 1855–1861 (2004)
12. Dixon, C., Frew, E.W.: Decentralized extremum-seeking control of nonholonomi vehicles to form a communication chain. Lecture Notes in Computer Science, vol. 369. Springer-Verlag (2007)
13. Dixon, C., Henkel, D., Frew, E.W., Brown, T.X.: Phase transitions for controlled mobility in wireless ad hoc networks. In: AIAA Guidance, Navigation, and Control Conference, Keystone, CO (2006)
14. EUROCONTROL/FAA: Future Communications Study Operational Concepts and Requirements Team: communications operating concept and requirements (COCR) for the future radio system. Tech. Rep. 1.0 (2006)
15. Fall, K.: A delay-tolerant network architecture for challenged internets. In: SIGCOMM '01, pp. 27–34 (2003)
16. Frew, E.W., Dixon, C., Elston, J., Argrow, B., Brown, T.X.: Networked communication, command, and control of an unmanned aircraft system. AIAA Journal of Aerospace Computing, Information, and Communication **5**(4), 84–107 (2008)
17. Harvey, D.J., Lu, T.F., Keller, M.A.: Comparing insect-inspired chemical plume tracking algorithms using a mobile robot. IEEE Trans. Robot. **24**(2), 307–317 (2008)
18. Henkel, D., Brown, T.X.: Optimizing the use of relays for link establishment in wireless networks. In: Proc. IEEE Wireless Communications and Networking Conference (WCNC), Hong Kong (2008a)
19. Henkel, D., Brown, T.X.: Towards autonomous data ferry route design through reinforcement learningi. In: Autonomic and Opportunistic Communications Workshop (2008b)
20. Henriksen, S.J.: Estimation of future communications bandwidth requirements for unmanned aircraft systems operating in the national airspace system. In: AIAA InfoTech@Aerospace, vol. 3, pp. 2746–2754. Rohnert Park, CA (2007)
21. How, J., King, E., Kuwata, Y.: Flight demonstrations of cooperative control for uav teams. In: AIAA 3rd "Unmanned-Unlimited" Technical Conference, Workshop, and Exhibit, vol. 1, pp. 505–513. Chicago, IL (2004)
22. Jenkins, A., Henkel, D., Brown, T.X.: Sensor data collection through gateways in a highly mobile mesh network. In: IEEE Wireless Communications and Networking Conference, pp. 2786–2791. Hong Kong, China (2007)
23. Lamb, G.S., Stone, T.G.: Air combat command concept of operations for endurance unmanned aerial vehicles. Web page, http://www.fas.org/irp/doddir/usaf/conops_uav/ (1996)

24. MLB Company: MLB Company—The Bat. http://www.spyplanes.com/bat3.html (2008)
25. Mohammad, A.J., Frost, V., Zaghloul, S., Prescott, G., Braaten, D.: Multi-channel Iridium communication system for polar field experiments. In: International Geoscience and Remote Sensing Symposium (IGARSS), vol. 1, pp. 121–124. Anchorage, AK (2004)
26. NOAA: NOAA News Online: NOAA and partners conduct first successful unmanned aircraft hurricane observation by flying through Ophelia. http://www.noaanews.noaa.gov/stories2005/s2508.htm (2005)
27. Office of the Secratary of Defense: Unmanned Aircraft Systems Roadmap: 2005–2030 (2005)
28. Ryan, A., Xiao, X., Rathinam, S., Tisdale, J., Zennaro, M., Caveney, D., Sengupta, R., Hedrick, J.K.: A modular software infrastructure for distributed control of collaborating UAVs. In: AIAA Guidance, Navigation, and Control Conference, Keystone, CO (2006)
29. Sofge, E.: Houston cops test drone now in Iraq, operator says. Web page, http://www.popularmechanics.com/science/air_space/4234272.html (2008)
30. Taub, B., Schilling, D.L.: Principles of Communication Systems. McGraw-Hill, New York (1986)
31. Vaglienti, B., Hoag, R., Niculescu, M.: Piccolo systrem user's guide: software v2.0.4 with piccolo command center (pcc). http://www.cloudcaptech.com/resources_autopilots.shtm#downloads (2008)
32. Wagner, B.: Civilian market for unmanned aircraft struggles to take flight. In: National Defense Magazine (2007)
33. Weibel, R., Hansman, R.J.: Safety considerations for operation of unmanned aerial vehicles in the national airspace system. Tech. Rep. ICAT 2005 01 (2006)
34. Zajkowski, T., Dunagan, S., Eilers, J.: Small UAS communications mission. In: Eleventh Biennial USDA Forest Service Remote Sensing Applications Conference, Salt Lake City, UT (2006)

UAVs Integration in the SWIM Based Architecture for ATM

Nicolás Peña · David Scarlatti · Aníbal Ollero

Originally published in the Journal of Intelligent and Robotic Systems, Volume 54, Nos 1–3, 39–59.
© Springer Science + Business Media B.V. 2008

Abstract The System Wide Information Management (SWIM) approach has been conceived to overcome the capacity and flexibility limitations of the current ATM systems. On the other hand the commercial applications of Unmanned Aerial Vehicles (UAVs) require the integration of these vehicles in the ATM. From this perspective, the unavoidable modernization of the ATM is seen as an opportunity to integrate the UAVs with the rest of the air traffic. This paper is devoted to study the feasibility and impact of the aggregation of UAVs on the future ATM supported by a SWIM inspired architecture. Departing from the existing technical documents that describe the fundamentals of SWIM we have explored the compatibility with a potential UAVs integration and also explored how the UAVs could help to improve the future ATM system. We will use the weather application as an example in both cases.

Keywords UAV · Air traffic management · System wide information management

1 Motivation and Objectives

The number of aircrafts in operation all around the world has grown steadily from the hundreds to the tens of thousands. As technology advanced, new systems and services were added to the Air Traffic Management (ATM) to improve safety and capacity, but due to the lack of a standard procedure to insert new functionalities in the ATM, these systems were designed independently, with very different interfaces

N. Peña (✉) · A. Ollero
University of Seville, Robotics, Vision and Control Group,
Avd. de los Descubrimientos s/n, 41092, Sevilla, Spain
e-mail: nicolas.grvc@gmail.com

D. Scarlatti
Boeing Research & Technology Europe – Cañada Real de las Merinas, 1-3,
Building 4-4th floor, 28042 Madrid, Spain

and had to be hard wired to each other in a specific way for every combination that needed to interoperate. As the number of present systems grew, the cost of inserting a new one was always higher.

The result of this trend is that the current ATM is a rigidly configured, complex collection of independent systems interconnected by very different technologies from geographically dispersed facilities. Then, they are expensive to maintain, and their modifications are very costly and time consuming. Future capacity demands require the implementation of new network-enabled operational capabilities that are not feasible within the current ATM systems. In fact, the safety, capacity, efficiency and security requirements to meet the expected demand require the application of new flexible ATM architectures. A new approach to face this future demand is the so-called System Wide Information Management (SWIM) [1, 2]. This system enables shared information across existing disparate systems for network-enabled operations, and improves air traffic operations by integrating systems for optimal performance [2, 4].

The commercial applications of Unmanned Aerial Vehicles (UAVs) require the integration of these vehicles in the ATM [5]. Currently, the UAVs operate in a completely segregated aerial space due to the absence of protocols for their integration in the Air Traffic Management systems. From this perspective, the planned modernization of the ATM is seen as an opportunity to integrate the UAVs with the rest of the air traffic in a single aerial space. In fact the implementation of the SWIM concept makes easier the integration of the UAVs in the ATM than the architecture in use. Moreover, the standardization of interfaces and the involved network centric concepts involved in SWIM are additional benefits for the integration of UAVs.

This paper is devoted to study the impact of the aggregation of UAVs on the future ATM supported by a SWIM inspired architecture. There are some publications and on going projects on the UASs integration in the aerospace. These usually focus the aspects that need to be improved before this integration happens such as the autonomous sense and avoid [6] or the safety requirements [7]. In [8] a new architecture for this integration is proposed.

This paper studies the integration of the UAVs in the ATM from the point of view of the actual plans for the future SWIM inspired, ATM [2, 3]. The effect at the different layers of the planned ATM structure is discussed. The paper will also explore the possibility that arises from the integration of the UAVs in the ATM that can be achieved in several layers of the proposed architecture. For example, regarding the network layer, which is heavily stressed by the ever growing aircrafts density, it is possible to consider stratospheric UAVs providing in a dynamic way additional bandwidth in areas with such requirements. Another example, in the application layer, is the service provided by the weather application that could be improved by UAVs acquiring information from areas with higher uncertainty on weather conditions.

On the other hand, for a proper integration in SWIM, the software and hardware architectures of the UAVs should be adapted. In terms of the hardware on-board, the limited payload should be considered to prioritize (in terms of safety) the on-board equipment included in the UAV. For instance, the ACAS (Airborne Collision Avoidance System) system should be included and integrated in the software on-board to allow an automated response for collision avoidance in the same way that manned aircrafts do.

We have centered our study in two general aspects of this integration. One is at the application level, where the functionality of the different services as surveillance and weather is offered to the SWIM clients. The other one looks at the layers below the application one to try to evaluate if the approaches proposed for some of the inners of the future ATM, like the proposed client server model and the data models, would be a problem or an advantage for the integration of the UAVs. The paper contains a section regarding to each of the two aspects studied by the authors.

2 Application Layer

This section presents some considerations regarding the SWIM application layer that are useful to put in context some concepts that will be presented later on.

The integration of the UAVs and their associated infrastructure in SWIM applications will be studied from two different approaches:

- UAVs using SWIM applications during their operation, and
- UAVs providing services intended to improve the performance of some SWIM applications.

Regarding the first approach, a common application, such as the weather application, will be selected in order to describe how an UAV will use the weather services. Thus, the components interacting with the UAV will be considered and the differences with respect to a conventional aircraft at the application level (i.e. automated periodic weather reporting instead of on pilot demand), as well as the requirements for the UAVs (on-board probabilistic local weather model, permanent link with a broker, required bandwidth, etc) will be examined. Figure 1 shows the proposed architecture of the weather application in a SWIM enabled NAS (National Airspace System). The elements shown in the Figure will be used in the rest of this section in a simplified manner referring to them as weather data producers, weather data repositories, weather brokers and finally, weather application clients.

The second approach considers UAVs providing an improvement in the applications of SWIM and even new services for manned aircrafts, UAVs and all clients in general. Some examples could be: provide weather information from locations with high uncertainty, response in emergencies, serve as flying repositories of pseudo-static information such as mid and long-term weather information, etc.

In this section, the weather application will be taken as a case study in order to clarify some aspects related to the integration of the UAVs in SWIM.

2.1 UAVs Using SWIM Applications During their Operation

In the application layer, it is relevant to distinguish between UAVs with minor human intervention and UAVs with a remote human pilot or operator. In this section, the analysis will be focused on the first one due to the following reasons:

- When the UAV is teleoperated, the human operator can play the role of a conventional pilot in terms of using the SWIM applications (i.e. processing the messages from a SWIM weather application). Therefore, in this case and from

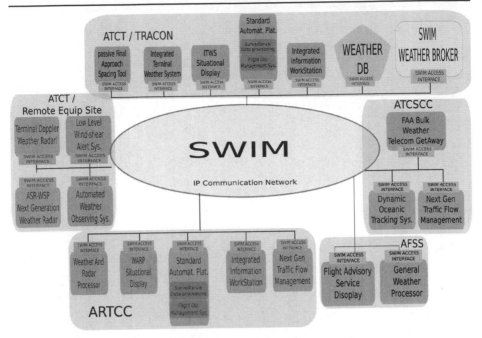

Fig. 1 The weather application elements adapted to the SWIM concepts

the point of view of the application layer, there is no relevant difference between an UAV and any other conventional aircraft. As it will be discussed later, in other layers several differences arise.

- Although nowadays almost all UAVs require in some extend the intervention of a human teleoperator, it is expected a transition towards full autonomy [9], allowing one operator to manage several UAVs. Furthermore, full autonomy would be even possible in some SWIM applications as weather remote sensing using stratospheric UAVs [10], reducing their operational cost.

Then, in the following, only autonomous UAVs will be considered. Therefore, in the weather application for example, problems related to autonomous weather messages processing and decision making should be addressed.

In the current ATM systems, after receiving one message, the human pilot has to "decode" and process it in order to make a decision taking into account other parameters such as the type of aircraft, payload, remaining fuel, etc. METAR (Meteorological Terminal Aviation Routine Weather Report) and TAF (Terminal Aerodrome Forecast) information are essential for flight planning and in-flight decisions. TAF messages are a very concise, coded 24-hour forecast for a specific airport that, opposed to a public weather forecast, only addresses weather elements critical to aviation as wind, visibility, weather and sky condition. A more recent type of message is the Transcribed Weather Enroute Broadcasts (TWEBs) which are composed by some Weather Forecast Offices (WFOs) and contain very similar information to a TAF but for a 50 mile wide corridor between two or three frequently connected airports.

As far as an autonomous UAV should perform this whole message processing by itself, the following requirements could be considered for the decisional level of an UAV:

- Estimate when a weather report is required. If the uncertainty about the weather in a given area of the flight plan is higher than a given threshold, the UAV should ask for a weather report in order to reduce the uncertainty of its local weather model. The weather reports requests will have a set of parameters such as the area of interest for example. Although a periodic weather reports request scheme could be adopted, the capability to autonomously estimate when a report is required will decrease the limited available data bandwidth.
- Process and decode the standard weather messages formats. Some examples of those formats are shown in Fig. 2.

In the proposed SWIM data models exposed in [3], several approaches are possible regarding the weather messages:

(a) Embedding the formats currently used into SWIM weather messages: The UAV should have to decode and process those formats.

There are several software projects which deal with METAR and TAF messages and even web sites providing access to some of these messages in a human readable format as [11].

In particular, the metaf2xml software [12] parses and decodes aviation routine weather reports and aerodrome forecasts (i.e. METAR and TAF messages) and stores the components in XML. They can then be converted to plain language, or other formats (see an example in Fig. 3). Similar software could be running on board the UAV, parsing the SWIM weather messages. After the parsing process, the information should be used to update the local weather model. A similar operation is performed by a software module of the FlightGear project [13] which updates a global weather model from parsed weather messages.

```
SBGL 091800Z 14000KT 9999 FEW020 BKN035 23/15 Q1013

SBGL 091550Z 091818 20015KT 8000 BKN020 PROB40 2024 4000 RA BKN015 BECMG 0002 24005KT

  BKN015 TEMPO 0210 4000 RA BR BKN010 BECMG 1113 20010KT 4000 RA BR TX22/19Z TN18/08Z

KJFK 091751Z 34013KT 10SM SCT038 21/11 A2952

KJFK 091425Z 091412 34012KT P6SM OVC025 TEMPO 1416 SCT025 FM1600 32013G18KT P6SM
```

```
  SCT025 BKN040 FM1800 31010KT P6SM SCT050 FM2200 28010KT P6SM SKC

RJTT 091830Z 34005KT CAVOK 14/09 Q1020 RMK A3012

RJTT 091500Z 091524 30004KT 9999 FEW020

RJTT 091500Z 100018 05005KT 9999 FEW030 BECMG 0003 14006KT BECMG 0912 30010KT
```

Fig. 2 Some METAR and TAF messages from Rio (SBGL), New York (KJFK), and Tokyo (RJTT)

msg: SBGL 091800Z 14008KT 9999 FEW020 BKN035 23/15 Q1013

METAR	METAR Report		
SBGL	Airport-Id:	SBGL	
091800Z	Report time:	on the 9., 18:00 UTC	
14008KT	Wind:	from the SE (140°) at 14.8 km/h	8 kt = 9.2 mph = 4.1 m/s
9999	Visibility:	>=10 km	>=6.2 US-miles
FEW020 BKN035	ceiling:	at 3500 ft	1070 m
	Sky condition:	few clouds at 2000 ft	610 m
		broken clouds at 3500 ft	1070 m
23/15	Temperature:	23 °C	73.4 °F
	Dewpoint:	15 °C	59 °F
	relative humidity:	61%	
Q1013	Pressure:	1013 hPa	29.91 in. Hg

Fig. 3 Example of an automatic parsing and conversion to HTML of a METAR message

(b) Changing the formats currently used: One of the advantages of the proposed network-centric brokers based SWIM architecture is that allows fusing information from several sources to generate a response for a given request. The message with the response can have a new unified SWIM format designed to be easily decoded and processed. In fact, as the messages are generated by a broker merging several sources of information, the UAVs themselves only need to provide basic weather reports and a complex local weather model is not required.

   The approach (a) would allow an easier and faster adaptation to SWIM (not for the UAVs integration in SWIM), but (b) takes full advantage of the SWIM architecture to provide better weather services.

*   Autonomous decision making integrating the weather information with other parameters such as UAV dynamics, payload, remaining fuel, relevance of its mission, etc. Resulting plans should be directly reported with a proper data structure. In any case, changes in data structures reported periodically such the planned 4D trajectory, would also inform about the new local plan to other SWIM entities subscribed.

   Regarding the UAV's autonomous decision functionality, it would be mandatory to periodically check that this mechanism is working properly. For example, the local plan or 4D trajectories reports could include the list of identifiers of weather messages used by the autonomous decision software. Furthermore, the decision making mechanism could be replicated in a ground facility in order to detect failures in the UAV operation (UAV malfunction), or eventually to protect against malicious intervention. In such a case, a human operator could take the control over the UAV through SWIM specialized channels that offer the CoS needed for teleoperation.

Finally, in order to illustrate an example of an UAV using the weather application over SWIM, the following storyboard based on Section 8 of [3] is presented. In a first stage (see Fig. 4), a weather database is updated with the messages sent from the weather data sources connected to SWIM. Following the publish/subscribe paradigm, a broker is acting as an intermediate element in the transaction, decoupling the data sources from the database itself.

In a second step, an UAV detects that its local weather model has a high level of uncertainty along its route and requests additional information. This query is managed by the nearest SWIM broker and sent to a weather data repository, where it is interpreted and processed (see Fig. 5). Another possible and more complex approach could involve a broker with the capability to process the query and translate it into a set of simple requests for different weather repositories. The two options differ in that the second one requires more intelligence in the brokers and less in the Data Repositories, taking the brokers a more than 'middle man' role. While the later can be seen as a more complex approach, it is more in line with the global SWIM philosophy of a really distributed, network centric system where data brokers for each service play a key role, isolating as much as possible the service clients from the inners of the backend elements of the service and the other way around.

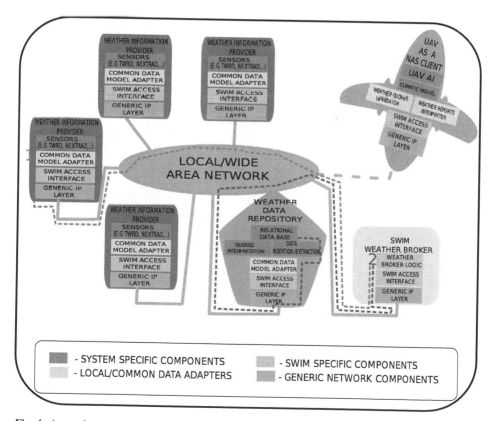

Fig. 4 A weather sensor connected to NAS and adapted according to SWIM reporting data. The information is sent to a database repository (or many) following the publish/subscribe paradigm

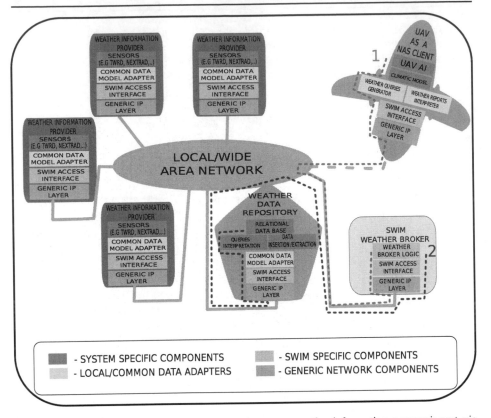

Fig. 5 When the UAV decision making core requires more weather information, a query is sent, via the closer broker, to a weather data repository

In the final stage (see Fig. 6), the required information is generated by the weather repository, formatted to be processed by the UAV and sent in a message to the broker, which just forwards it to the UAV. The message formatting task could have been also performed by the broker. Actually this would be obligatory in the ideal case of more intelligent brokers commented before. In any case, this would have been transparent for the rest of SWIM elements.

Once the response has arrived to the UAV the new weather information would be finally inserted in the local weather model satisfying the need that initially caused the request.

2.2 UAVs Improving the Performance of SWIM Applications

The adoption of the SWIM concept provides a unique opportunity to improve existing services and provide new ones. The current architecture has independent subsystems and services, and different physical connections for each type of information, whereas the SWIM architecture is network-centric and designed to grow and adapt to future demands. Scalability and connectivity are inherent features in SWIM and the cost of adding new services is reduced when comparing with the current architecture [14].

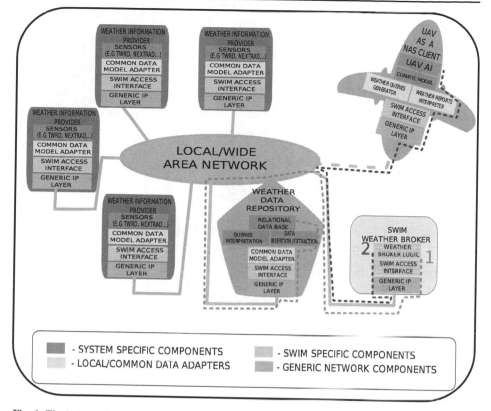

Fig. 6 The information requested is sent to the UAV

Taking into account the inherent scalability of SWIM and the emerging UAV technologies, it is expected to have UAVs supporting SWIM applications and even new UAV based applications. In the next subsections, several examples are provided assuming a flexible SWIM architecture with data fusion capabilities consistent with the COP concept [3]. Moreover, it is assumed that the brokers can manage "abstract" requests (not only forwarding raw data provided by sensors).

2.2.1 UAVs Acting as Weather Sensors

A team of stratospheric UAVs could be used to gather weather information [10] from areas with a weather estimation uncertainty higher than a given threshold. The autonomy of this type of UAVs is being improved by the recent developments in photovoltaic technology with the goal of achieving unlimited autonomy [15].

If it is expected to have traffic in a zone with high weather uncertainty, the weather servers could compute a list of waypoints to be visited for gathering weather data. Those servers would have access to the future traffic requesting this information to the corresponding brokers. Therefore, the waypoints could be prioritized depending on the expected routes in this zone.

The waypoints list could be sent to a stratospheric UAVs team that will autonomously allocate the waypoints among themselves using distributed algorithms trying to optimize some criteria, such as minimizing the total mission time or cost.

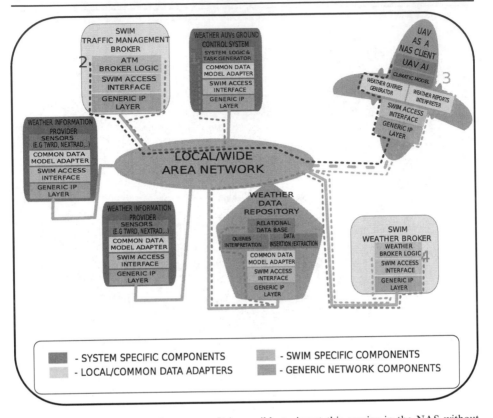

Fig. 7 UAV acting as a weather sensor. It is possible to insert this service in the NAS without changing the SWIM design

Figure 7 shows how the UAVs, acting as weather data sources, fit well in the proposed SWIM architecture. In this figure, four stages have been identified by different types of connecting lines in the information flow. In the first phase, composed by two stages, the UAV receives (via an ATM broker) a request consisting in a "goto" command. The UAV executes it and go to the specified location to gather weather information with the sensors on-board. In the second phase, the UAV reports weather data to the closest SWIM weather broker and then send the information to the weather repository.

2.2.2 Surveillance in Specific Locations and Emergency Response

Nowadays, UAVs are mainly used in surveillance missions taking aerial images from a given area. Those missions are mainly for military purposes [9], but some civil applications can also be found [16, 17]. In the future, those surveillance functionalities could be integrated in SWIM providing services such as autonomous surveying of a disaster area or assistance to identify aircrafts present in the surveillance system that are not responding to the radio.

There are several companies that provide satellite images of specific areas during a given period. Low orbit satellites can provide sequences of images during a period of time limited by their orbital speed. Those companies could adopt the use of UAVs to provide more flexible services at a lower cost. Their clients could even have access in real time to the images via a web browser in a computer connected to SWIM. A quote from [18]: "The progressive implementation of the SWIM principles in AIM (Aeronautical Information Management) is in fact AIM's evolution to IM, or Information Management that is fully SWIM based and which is the ultimate goal".

A fully SWIM compatible system could be easily interconnected with other network-centric platforms allowing to increase the number of services provided. Furthermore, as far as the amount of shared information and resources will be increased, the cost of implementing new services will decrease.

In order to illustrate an application that could exploit the benefits of interconnecting SWIM to other existing networks, the following situation will be considered, as a generalization of the application of UAVs for fire monitoring [19]. A fire station in Madrid receives a fire alarm from the sensors located in a given building. Then, an autonomous visual confirmation process starts sending a request to GIM (General Information Management) for images from the building area. GIM is part of a global network integrated by many networks including SWIM and therefore the request is finally routed to a SWIM broker. Several manned and unmanned aircrafts are flying over the city and the broker selects a proper one equipped with a camera on-board to establish a direct link between the camera and the fire station. The camera pan&tilt is pointed to the building area and the video streaming is received in the fire station allowing confirm/discard the fire (this streaming is allowed in the third broker model presented in [3]).

3 Middleware and the Network-Centric Nature

3.1 Introduction

In this section, the impact of the UAVs insertion in the SWIM middleware is studied. As far as SWIM is being designed to be very scalable and generic with the stated goal of easing the integration of new elements, the following analysis can be considered as an indicator that it is well designed for it, at least in the UAVs case.

In the current SWIM specification, it is described how piloted aircrafts (or any other data source/sink) can publish their information and subscribe to any authorized data channel. In this section it will be analyzed if those procedures would also allow the integration of UAVs in a transparent way for other SWIM users. For this purpose, the main SWIM architecture concepts are revised in the next subsection to check if they are general enough for this seamless integration. In the next subsections, required changes or additions for the UAVs integration that had been spotted during our analysis are listed and explained.

3.2 SWIM Main Concepts and UAVs

3.2.1 Network-centric Nature

One of the main ideas present in SWIM is to connect all the systems integrating ATM in a uniform manner with well defined common interfaces, all connected to a network with information flowing from the sources to the sinks sharing generic routing channels. This information can be also processed in any intermediate subsystem

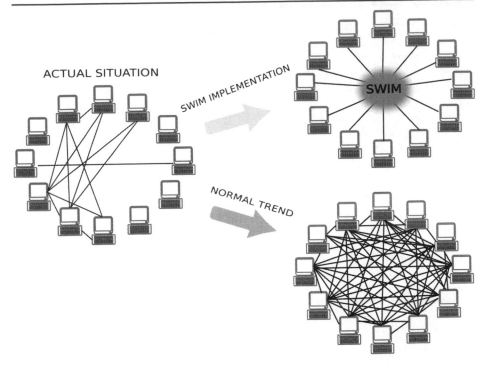

Fig. 8 SWIM network-centric approach versus point to point dedicated data link connections

demanding it. This concept represents a real revolution in the civil aviation as far as nowadays, ATM is mainly composed by many independent subsystems connected by dedicated channels in a very rigid way (see Fig. 8). Therefore, any change such as inserting a new subsystem has a significant associated cost.

On the other hand, in the recent research area of multi-UAV systems the architecture adopted has usually been network-oriented as it is the natural solution that leads to a flexible and cost effective interconnection among multiple systems. Regarding the physical communication layer, it is possible to find solutions with different channels for different types of information, i.e. high bandwidth analog channels for images transmission [20] or dedicated channels for teleoperation [21]. But the principal trend nowadays is to have one unique digital channel (or more if fault tolerance is a requirement) shared by different types of information. The progress in telecommunications technology is making this approach feasible.

Therefore, as the recent developments in multi-UAV systems follow a network-centric approach close to the SWIM architecture, from this point of view the introduction of the UAVs in SWIM is possible and even can be considered as natural.

3.2.2 The Publish/Subscribe Paradigm

Once a network-centric approach is adopted, the elements connected to the network can interchange information according to different models. Some of them are listed in Table 1, extracted from [3], which compares them with respect to the degree of decoupling between data producers and consumers.

Table 1 Different data distribution models compared w.r.t the degree of decoupling between information producers and consumers

Abstraction	Space decoupling?	Time decoupling?	Synchronization decoupling?
Message Passing	No	No	Publisher side
RPC/RMI	No	No	Publisher side
Async. RPC /RMI	No	No	YES
Notifications	No	No	YES
Shared Spaces	Yes	Yes	Publisher side
Message Queuing	Yes	Yes	Publisher side
Publish Subscribe	Yes	Yes	Yes

A high degree of decoupling leads to more robust solutions and to lower costs for the adaptation or insertion of new elements, which are properties required in SWIM. In general, the publish/subscribe paradigm allows a total decoupling between the sources and the sinks of information. Furthermore, new sources or sinks can be added or removed dynamically with a minor impact on the operation of the network. In fact, the publish/subscribe paradigm has been already adopted in other areas such as wireless sensor networks or multi-robot research [17, 22, 23], where the properties mentioned above are also relevant. Moreover, the performance of the network scales well against changes in the demand of a given type of information if dynamic replication techniques are applied.

In the publish/subscribe paradigm, the flow of information between sources and sinks is managed dynamically by one or several intermediate entities, usually called data brokers (see Fig. 9). The middleware is composed by these intermediate entities and their communication protocols. As a result, there is no direct communication between sources and sinks and if a new subsystem has to be added to the network, it is only necessary to develop its interface with the data brokers. Therefore, compatibility tests between this subsystem and other existing subsystems in the network are not necessary, decreasing the integration cost and time. In the design of SWIM, the publish/subscribe architecture was adopted from the beginning to allow an easy integration of new subsystems in the NAS, such as the UAVs themselves.

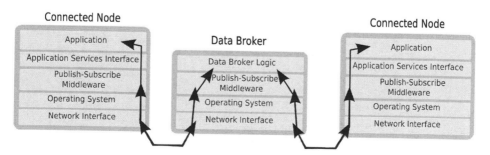

Fig. 9 Pure publish/subscribe model, with no direct interaction between clients

3.2.3 Architectures for the Brokers

There are several possible architectures for the data brokers, depending on the required services and capabilities for the whole communication system. Form the architectural point of view, the simplest one could be similar to the solution adopted in [17], where there is no independent element managing the information exchanges. The implementation of the middleware interface in the clients provides this functionality in a distributed way as shown in Fig. 10. This architecture does not follow a pure publish/subscribe paradigm, but allows low latencies and it is a good solution when real time operation is required.

On the other hand, the most complex architecture could use different protocols to offer pure publish/subscribe functionality for services requiring low bandwidth and a direct connection oriented protocol for streaming data services.

There are three different brokers architectures proposed for SWIM in [3] and all the information related to the SWIM middleware design is consistent with the analysis provided in this paper. Table 2, extracted from the executive summary of [3], shows the main differences between the architectures proposed for the SWIM brokers. It should be pointed out that, due to the network centric nature of SWIM, more than one model can operate at the same time in the same network depending on the application and data types involved in the communication.

On the other hand, the complexity of the communication system, and hence the fault probability, would be increased, at least linearly, with the number of different broker models implemented. Therefore, this number should be kept as low as possible and the integration of the UAVs services should be adapted to the models proposed in Table 2, which are general enough:

- Model "Pub/Sub Broker": It follows the strict publish/subscribe paradigm, so the UAVs will only have to communicate with the brokers and the integration would be easier. The latency associated with this model makes it incompatible with some of the most common UAV applications nowadays, such as teleoperation. Furthermore, there are also services offered by the UAVs such as surveillance video streaming that generates a massive amount of information and the use of data brokers could represent a bottleneck, increasing the latency and decreasing the quality of service.

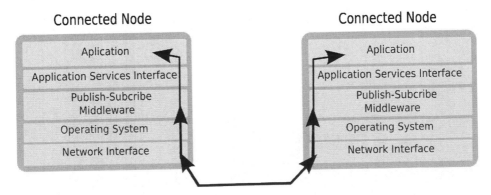

Fig. 10 Distributed version of the publish/subscribe paradigm

Table 2 Three different architectures proposed for the SWIM brokers

Broker Model	Description	Advantages	Disadvantages
Pub/Sub Broker	Implements the full set of functions needed to offer a pure Pub/sub middleware so that all operations between publishers and subscribers are completely decoupled in time, space and synchronization	Changes in publishers or subscribers completely transparent to each other; unified management of exchanged information	Possible extra latency in the process as data flows through the broker; This fact also means that the broker is a potential bottle neck limiting the ability to handle large streams of data
Lightweight Broker	Implements a subset of the functions needed to offer a pure Pub/sub middleware and implements the rest using traditional messaging mechanisms. This mixture means that not all operations between publishers and subscribers are completely decoupled	Supports implementation of several variants of Pub/Sub schemes	Publishers and Subscribers are not fully decoupled; predefined data channels need to be established
VC Broker	VC Broker is a superset of the functionalities offered by the pure Pub/Sub model. It can offer completely decoupled communication between the endpoints but also implements the primitives needed to establish a Virtual Circuit between to of them to achieve better latency or higher bandwidth	Broker approach is tailored to data type; Can offer different QoS for different data types	Extra complexity in information management functions such as monitoring; When connected by virtual circuits the Publishers and Subscribers are not fully decoupled

- Model "Lightweight Broker": This model is intended to simplify the adoption of SWIM by using more classical approaches that would allow reusing some of the current interfaces. But the integration of the UAVs would require to implement those interfaces as special cases that do not follow the publish/subscribe policy implemented in the brokers. This option can be cost effective in the short term, but it is not as uniform or powerful as the other two models.
- Model "VC Broker": In this solution, the broker offers a different kind of service depending on the data type. It works as the first model when it is enough to fulfill the service requirements. But when low latency or high bandwidth is required, the broker provides a virtual connection between the data producer and consumer to prevent the potential bottleneck due to centralized brokers. This model is general enough to fulfill all the requirements from the point of view of the UAVs integration. For most of the services, the UAV could use the

publish and subscribe pure model, as any other SWIM client. But for latency sensible applications such as teleoperation [24] or high bandwidth requirements as real time surveillance video transmission, virtual circuits between the UAV and the client application can be created by the broker dynamically.

3.2.4 Access Solutions Proposed for the SWIM Clients

Figure 11 shows three different models proposed to connect to SWIM. In the first model there is a SWIM interface software which is running on the hardware on-board and interacting with the brokers. This model is preferred for the UAVs integration as far as it is the most flexible solution in terms of the potential spectrum of services that can be offered and, additionally does not require increasing the payload. The downside of this solution is that the interface is not so decoupled from the UAV specific software and hardware as in the other options, and special tests could be required to check the implementation and performance of the SWIM interface in order to avoid security problems.

The second model is based on specific hardware to support the connection of currently NAS integrated subsystems to SWIM. Those subsystems have specific interfaces that require a hardware adaptation (see Section 8 in [3]). This model is not necessary for the UAVs as far as the hardware on board is quite flexible and updatable. In fact, the next generation of UAVs could be designed with SWIM hardware compatibility. Anyway, this model allows addressing the security issues much better than the others because the SWIM interface is based on a specific hardware that can be designed following "trusted computing" principles.

Finally, the third model is based on standard web browsers, whose services can be "upgraded" by new web 2.0 technologies in the near future. In the last years, web browsers have been used for teleoperation applications in robotics and even in some

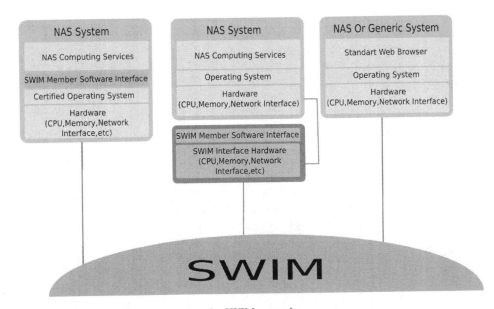

Fig. 11 Different models to connect to the SWIM network

UAVs control centers. Therefore, it is possible to have SWIM clients based on web browsers providing access to the UAV services and, in the opposite direction, the UAVs could also have access to information servers, such as map servers, through a web interface. In any case, limitations due to the latency and associated overhead involved in this model should be considered in order to select which interface to use for each service provided or required. The advantage of using this third model is that it is easier to develop and that the web browser/server protocols are well tested and designed with security in mind.

A combination of the first and third models (for non critical data queries) could be the best option.

3.2.5 UAVs Interfaces for Data and Services Access

The SWIM data model is described in the Appendix G of [3], where a detailed study of the different factors to be taken into account in the specification of the SWIM data structures is provided. Moreover, the impact of the data model on the flexibility and complexity of adding new subsystems is also presented.

Regarding the data model, the main design decisions made up to now and found in the used references can be summarized as follows:

- All the information should have a unified format, which has been called "SWIM Common Data Model" (see Fig. 12). Therefore, all the messages must be embedded in a standard data container, whose header should include standard fields independent of the type of message, such as its source, the data time stamp, etc. This mandatory header required in all the SWIM messages contains a set of fields which are usually referred as "SWIM Common Data Fields". During the development of SWIM it is expected to have sub-headers corresponding to sub-classes of data types leading to a tree of standard data structures. For every type of message, this tree will contain the list of mandatory fields and their associated

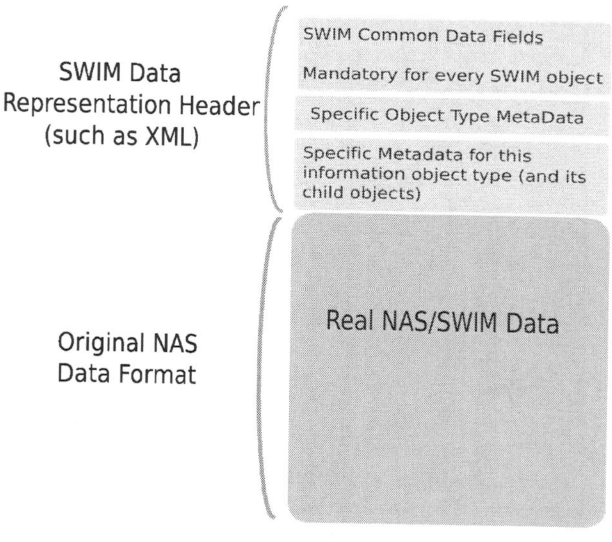

Fig. 12 Common data model diagram

SWIM Data Representation Header (such as XML)

SWIM Common Data Fields

Mandatory for every SWIM object

Specific Object Type MetaData

Specific Metadata for this information object type (and its child objects)

Original NAS Data Format

Real NAS/SWIM Data

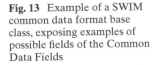

Fig. 13 Example of a SWIM
common data format base
class, exposing examples of
possible fields of the Common
Data Fields

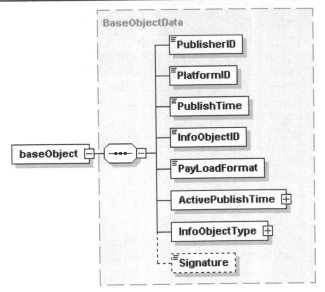

data structures. An example of a SWIM common data format base class is shown in Fig. 13.

- The message structure should be embedded in the messages themselves, allowing a client to process a message without any previous knowledge about its internal data structure. This feature can be implemented using metalanguages or markup languages, such as XML (eXtensible Markup Language), and provides flexibility in the messages generation. For example, if a field is not applicable in a given message, this field is not included. Other advantage when using markup languages is the low probability for bad interpreted messages. When a new message is received, only the fields with known identifiers are processed whereas the rest of fields are ignored. This characteristic also makes easier the migration from an old type of message to a new one. During the transition both the old and the new fields are sent in order to keep offering the same services.

- In an initial stage, the NAS messages in their current format should be embedded into the new SWIM messages to allow an easier migration of the existing subsystems. In fact, once the original NAS message has been extracted from the new message, the rest of the process will remain the same. Later, those old data structures can be fully substituted for XML messages.

Almost all the decisions summarized above makes easier the integration in SWIM of the UAVs, as any other kind of new NAS clients, due to the fact that they were adopted with the flexibility criteria in mind. Of course, this statement does not apply to the last point, which is only oriented to decrease the migration cost.

3.3 Aspects Requiring some Adaptation

In the previous subsection, it has been mentioned that in contrast to the rigid nature of the current NAS architecture, the flexible network-centric design of SWIM allows an easy integration of the UAVs. But there are several aspects that could require further considerations due to the particular characteristics of the UAVs. It should be pointed out that the required adaptations presented in the next subsections do not come from the autonomous nature of the UAVs. In fact, the software on-board can make this autonomy characteristic nearly transparent for the rest of NAS subsystems.

In the following subsections, several aspects that could require some adaptation to tackle with some particular characteristics of the UAVs are presented. Moreover, possible solutions built on top of the basic functionality provided by SWIM are also depicted. Those solutions are based on elements that could be also added to the current SWIM specification in order to provide services requiring high flexibility.

3.3.1 Dynamic Brokers

The current UAV applications usually involve taking-off and landing from temporal or improvised locations. From the proposed architecture for the hierarchical connection and distribution of SWIM brokers [3], it seems that the idea is to have at least one broker in every airport that manages the insertion of new aircrafts into SWIM.

If the integration process of the UAVs implies to operate from conventional airports, their functionally will be drastically reduced. Therefore, an UAV taking-off from a given location needs a procedure to signal that it has started to operate and requires SWIM services (even if it is far away from an airport) such as receiving the "clear for take-off" message. The whole procedure requires a connection to a broker.

Nowadays, the ground control stations of the UAVs are continuously connected to them, and in some aspects, act as airports control towers (ATCT) for the UAVs. Then, it seems a natural solution to equip the ground control station with a broker that can link the UAV with other NAS elements. This solution involves dynamic brokers that are subscribing and unsubscribing to SWIM continuously from different locations.

3.3.2 High Bandwidth Channels on Demand

The teleoperation of the UAVs or some specific UAVs services such as surveillance could require the transmission of high bandwidth data in real-time during certain periods. Therefore, protocols to establish dedicated communication channels on demand are also required.

3.3.3 Collaborative UAVs Surveillance System

Regarding surveillance, due to the small dimensions and furtive nature of some UAVs, all the UAVs should be equipped with GPS and should continuously broadcast their positions at a given rate. This positional information could be merged with the rest of the surveillance related information (as the one provided by the primary and secondary radars) by the related data brokers. Given that the GPS infrastructure has been improved in recent years with the goal of making it more reliable and useful for the FAA this possibility should be easy to implement.

New elements designed to increase the usefulness of the GPS system for ATM applications are:

- The Wide Area Augmentation System (WAAS) [24]: Created by the Federal Aviation Administration (FAA) to augment GPS with additional signals for increasing the reliability, integrity, accuracy and availability of GPS for aviation users in the United States.

- The Local Area Augmentation System (LAAS): Created to allow GPS to be used for landing airplanes. LAAS is installed at individual airports and is effective over just a short range. This system should help the autonomous landing and take off of UAVs in normal airports.

3.3.4 Special Communication Technologies

As far as satellite links are intended to be used for global coverage in SWIM (Satellite enhanced CNS), the payload and budget limitations of some types of UAVs in terms of communication equipment on-board could be a limitation. In such cases, it could be the responsibility of the UAVs associated ground station to provide the communications link with its UAVs. As this is not the best, most general solution, it is the most straight forward one as the only link between the UAVs and the rest of the SWIM components would by its associated ground control station. This is coherent with the proposed figure of dynamic data brokers presented in Section 3.3.1.

4 Conclusions

The transition from the current rigidly configured ATM approach to the SWIM based ATM architecture will represent an important change in ATM concepts and procedures. This transition could represent a good opportunity to facilitate the introduction of the UAVs in non-segregated aerial spaces, which has been recognized as one of the main barriers for commercial UAV applications.

The UAV integration is examined in this paper at different layers of the ATM structure from global concepts, as the network centric nature and the publish/subscribe paradigm, to the particular interfaces, broker and data models required to implement SWIM.

Furthermore, the UAVs integration could also help to improve basic ATM services, such as the weather information, and to offer new services such as on demand surveillance.

Then, it can be seen that the required extensions to include UAVs in the air traffic management of non-segregated aerial spaces are minimal and compatible with the proposed SWIM based ATM architecture.

References

1. Boeing Technology: Phantom works. Advanced Air Technology Management. http://www.boeing.com/phantom/ast/61605_08swim.pdf. Accessed 24 October 2007
2. SWIM Program Overview: http://www.swim.gov (redirects to www.faa.gov). Accessed 9 May 2008
3. System-Wide Information Management (SWIM) Architecture and Requirements. CNS-ATM Task 17 Final Report, ITT Industries, Advanced Engineering and Sciences Division, 26 March 2004
4. Jin, J., Gilbert, T., Henriksen, S., Hung, J.: ATO-P (ASD 100)/ITT SWIM Architecture Development. CNS-ATM, 29 April 2004

5. Koeners, G.J.M., De Vries, M.F.L., Goossens, A.A.H.E., Tadema, J., Theunissen, E.: Exploring network enabled airspace integration functions for a UAV mission management station. 25th Digital Avionics Systems Conference, 2006 IEEE/AIAA, Oct. 2006, pp. 1–11
6. Carbone, C., Ciniglio, U., Corraro, F., Luongo, S.: A novel 3D geometric algorithm for aircraft autonomous collision avoidance. In: 45th IEEE Conference on Decision and Control (CDC'06), pp. 1580–1585. San Diego, California, December 2006
7. UAV safety issues for civil operations (USICO), FP5 Programme Reference: G4RD-CT-2002-00635
8. Le Tallec, C., Joulia, A.: IFATS an innovative future air transport system concept. In: 4th Eurocontrol Innovative Research Workshop, December 2005
9. UAV Roadmap 2005–2030 – Office of the Secretary of Defense, August 2005
10. Everaerts, J., Lewyckyj, N., Fransaer, D.: Pegasus: design of a stratospheric long endurance UAV system for remote sensing. In: Proceedings of the XXth ISPRS Congress, July 2004
11. Weather Information Interface: http://aviationweather.gov/. Accessed 9 April 2007
12. metaf2xml: convert METAR and TAF messages to XML. Project web site: http://metaf2xml.sourceforge.net/. Accessed 25 May 2006
13. FlightGear Flight Simulator Project Homepage: http://www.flightgear.org/. Accessed 5 May 2008
14. Meserole, C.: Global communications, navigation, & surveillance systems program – progress and plans. In: 5th Integrated CNS Technologies Conference & Workshop, May 2005
15. Romeo, G., Frulla, G.: HELIPLAT: high altitude very-long endurance solar powered UAV for telecommunication and earth observation applications. Aeronaut. J. **108**(1084), 277–293 (2004)
16. Merino, L., Caballero, F., Martínez-de Dios, J.R., Ollero, A.: Cooperative fire detection using unmanned aerial vehicles. In: Proceedings of the 2005 IEEE, IEEE International Conference on Robotics and Automation, Barcelona (Spain), April 2005
17. Ollero, A., Maza, I.: Multiple Heterogeneous Unmanned Aerial Vehicles. Springer Tracts on Advanced Robotics. Springer, Berlin (2007)
18. Aeronautical Information Management Strategy, V4.0. EUROCONTROL, Brussels, Belgium March 2006
19. Merino, L., Caballero, F., Martínez-de Dios, J.R., Ferruz, J., Ollero, A.: A cooperative perception system for multiple UAVs: application to automatic detection of forest fires. J. Field Robot **23**(3), 165–184 (2006)
20. Beard, R.W., Kingston, D., Quigley, M., Snyder, D., Christiansen, R., Johnson, W., McLain, T., Goodrich, M.: Autonomous vehicle technologies for small fixed-wing UAVs. J. Aerosp. Comput. Inform. Commun. **2**(1), 92–108 (2005)
21. Alcázar, J., Cuesta, F., Ollero, A., Nogales, C., López-Pichaco, F.: Teleoperación de helicópteros para monitorización aérea en COMETS (in Spanish). XXIV Jornadas de Automática (JA 2003), León (Spain), 10–12 Septiembre 2003
22. Sørensen, C.F., Wu, M., Sivaharan, T., et al.: A context-aware middleware for applications in mobile AdHoc Environments. In: Proceedings of the 2nd Workshop on Middleware for Pervasive and Ad-hoc Computing Table of Contents. Toronto (2004)
23. Soetens, H., Koninckx, P.: The real-time motion control core of the Orocos project Bruyninckx, Robotics and Automation. In: Proceedings. ICRA '03. (2003)
24. Lam, T.M., Mulder, M., van Paassen, M.M.: Collision avoidance in UAV tele-operation with time delay, conference on systems, man and cybernetics. ISIC. IEEE International. Montreal, October 2007
25. Loh, R., Wullschleger, V., Elrod, B., Lage, M., Haas, F.: The U.S. wide-area augmentation system (WAAS). Journal Navigation **42**(3), 435–465 (1995)

A Survey of UAS Technologies for Command, Control, and Communication (C3)

Richard S. Stansbury · Manan A. Vyas ·
Timothy A. Wilson

Originally published in the Journal of Intelligent and Robotic Systems, Volume 54, Nos 1–3, 61–78.
© Springer Science + Business Media B.V. 2008

Abstract The integration of unmanned aircraft systems (UAS) into the National Airspace System (NAS) presents many challenges including airworthiness certification. As an alternative to the time consuming process of modifying the Federal Aviation Regulations (FARs), guidance materials may be generated that apply existing airworthiness regulations toward UAS. This paper discusses research to assist in the development of such guidance material. The results of a technology survey of command, control, and communication (C3) technologies for UAS are presented. Technologies supporting both line-of-sight and beyond line-of-sight UAS operations are examined. For each, data link technologies, flight control, and air traffic control (ATC) coordination are considered. Existing protocols and standards for UAS and aircraft communication technologies are discussed. Finally, future work toward developing the guidance material is discussed.

Keywords Command, control, and communication (C3) ·
Unmanned aircraft systems (UAS) · Certification

R. S. Stansbury (✉) · T. A. Wilson
Department of Computer and Software Engineering,
Embry Riddle Aeronautical University,
Daytona Beach, FL 32114, USA
e-mail: stansbur@erau.edu

T. A. Wilson
e-mail: wilsonti@erau.edu

M. A. Vyas
Department of Aerospace Engineering,
Embry Riddle Aeronautical University,
Daytona Beach, FL 32114, USA
e-mail: vyas85a@erau.edu

K. P. Valavanis et al. (eds.), *Unmanned Aircraft Systems.* DOI: 10.1007/978-1-4020-9137-7_5

Abbreviations

ATC	air traffic control
BAMS	broad area maritime surveillance [24]
BLOS	radio frequency beyond line-of-sight
C2	command and control
C3	command, control, and communication
CoA	certificate of authorization
CFR	Code of Federal Regulations
DSA	detect, sense, and avoid
FAA	Federal Aviation Administration
FAR	Federal Aviation Regulations
HF	high frequency
ICAO	International Civil Aviation Organization [13]
IFR	Instrument Flight Rules
LOS	radio frequency line-of-sight
NAS	National Airspace System
NATO	North Atlantic Treaty Organization
STANAG	standardization agreement
SUAV	small UAV
TFR	temporary flight restriction
TUAV	tactical UAV
UA	unmanned aircraft
UAV	unmanned aerial vehicle
UAS	unmanned aircraft system
UCS	UAV Control System [22]
UHF	ultra high frequency
VHF	very high frequency
VSM	vehicle specific module [22]

1 Introduction

The integration of unmanned aircraft systems (UAS) into the US National Airspace System (NAS) is both a daunting and high priority task for the Federal Aviation Administration, manufacturers, and users. To modify the existing federal aviation regulations (FARs) under 14 CFR [1] to certify the airworthiness of a UAS, a tremendous effort is needed taking more time than the interested parties desire. An alternative approach is the development of guidance materials to interpret existing airworthiness standards upon manned aircraft systems that they may be applied to UAS.

The research presented in this paper assists the effort of this new approach. To interpret existing standards for UAS systems, a survey of existing UAS technologies must be performed, a system model must then be developed, and finally a regulatory gap analysis performed. This paper presents the technology survey of command, control, and communication (C3) systems on-board existing UAS.

1.1 Problem and Motivation

Before unmanned aircraft systems (UAS) are certified for general operation within the national airspace, airworthiness standards must be adopted. These standards may be derived through two possible approaches. The first approach would require the adoption of a new part to 14 CFR (also known as the Federal Aviation Regulations, or FARs). The time required to make such a change would be much greater than desired by the FAA, the US Military, UAS manufacturers, and UAS users.

The second option would be the provide guidance materials regarding the interpretation of the FARs with respect to the components of the unmanned aircraft system. This will likely be the approach taken. This research is meant to support its progress. A two part study is being conducted to analyze the technology associated with command, control, and communication of UAS, and from the technology survey determine issues and regulatory gaps that must be addressed.

The technology survey is essential to supporting the regulatory gap analysis. From an understanding of the technologics, it is then possible to examine the FARs and determine instances in which the existing regulation is not aligned with the state-of-the-art. These instances are the regulatory gaps that will be documented as the second phase of this project.

1.2 Approach

To assist the technology survey, a C3 system model shown in Fig. 1 was developed. The goal of this system model is to logically break the survey down into logical categories. For UAS operations, aircraft may operate within radio frequency line-of-sight, or beyond line-of-sight. Technologies and operating procedures related to command, control, and communication of UAS are divided into one of these two categories.

Fig. 1 C3 system model

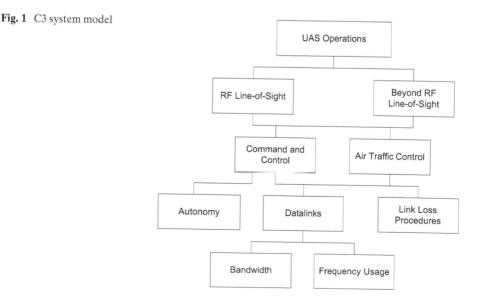

Under each category of RF LOS and BLOS, UAS technical issues may be divided into two categories: Command and Control (C2) and Air Traffic Control (ATC). For C2, the survey will explore technologies and issues necessary to safely support flight operations of the UAS from a remote pilot and/or control point-of-view. For ATC, technologies and issues related to the interaction of the aircraft or pilot-in-command with air traffic controllers while operating in the National Airspace System.

Under C2 and ATC, the various data links are examined including their respective frequency and data rates. The current link loss procedures are enumerated. Finally, for C2 only, the issue of autonomy, remote pilot versus autopilot, is examined for a variety of aircraft.

Communication related to detect, sense, and avoid (DSA) currently lies outside of the scope of this research. Such systems may consume additional command and control bandwidth, or require their own independent data links.

1.3 Paper Layout

Our system model divides our investigation of C3 technologies and issues between operation under RF line-of-sight and RF beyond line-of-sight conditions. The survey will discuss UAS operating under these conditions, the technological issues associated with command, control, and communication, and their interaction with ATC.

It is important to note that BLOS UAS do contain some LOS technologies. Figure 2 illustrates the overlap between these operating conditions and the class of UAS that can operate within these areas. The line-of-sight section shall include a discussion of all surveyed aircraft, but the beyond line-of-sight section shall only discuss medium and high-endurance UAS capable of operating beyond the RF line-of-sight of the pilot-in-command.

After discussing the surveyed UAS, security issues of C3 data links and flight controls are considered. Existing communication protocols, message sets, and standards are considered with respect to UAS C3.

For the language of this document, inline with current FAA documents, the term unmanned aircraft system will be used to describe the unmanned aircraft and its ground station. The term unmanned aerial vehicle, or its acronym UAV will be used only during discussion protocols in which the previous terminology is used.

Fig. 2 BLOS operations
subset of LOS operations

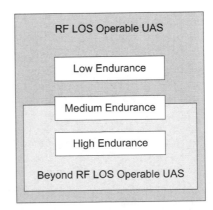

2 RF Line-of-Sight C3 Technologies and Operations

Line-of-sight operation may be divided between three classes of unmanned aircraft, which are low endurance, medium endurance, and high endurance. The first class operates almost entirely in line-of-sight. Surveyed low endurance aircraft include Advance Ceramic Research's Manta B [3], Advance Ceramic Research's Silver Fox [3], Meridian [10], Aerovironment's Raven [4], and Aerovironment's Dragon Eye [4]. The second and third class operates in both line-of-sight and beyond line-of-sight conditions. Surveyed medium endurance UA include Insitu's ScanEagle [15], Insitu's Georanger [15] and AAI Corp.'s Shadow [2], while the high endurance UA include General Atomics' Predator [7], General Atomics' Mariner [7], Northrup Grumman's Global Hawk [24], Northrup Grumman's BAMS [24], and Aerovironment's Global Observer [4]. Table 1 lists some examples of LOS C3 Technologies.

2.1 Data Links

The literature research revealed that LOS command and control data links commonly use a variety of frequencies from Very High Frequency (35 MHz) to C Band (6 GHz) [23]. It was observed from the technology survey that the most common LOS data link employed for current UAS is C Band. C Band uses low GHz frequencies for downlink, 3.7–4.2 GHz, and 5.9–6.4 for uplink [28]. C Band is strategically chosen for LOS C2 because the low GHz frequencies are less affected by extreme weather condition. For example, Mariner, Predator, and Predator's civilian versions such as Altair [5] use C Band for LOS C2.

Some small UA like ScanEagle and Georanger [15], Meridian [10], Shadow [2], Dragon [4], and Raven [4] use UHF for LOS command and control. It is not uncommon for these aircraft to utilize 72 MHz hand-held remote control similar or identical to those used by hobbiests.

Another option explored is Common Data Link (CDL) or Tactical CDL. CDL is a jam resistant spread spectrum digital microwave link only used by the military [8]. CDL is mostly used for BLOS operations; however, it can be used for LOS operations to ensure continuously safe and seamless communication when deployed in hostile territory. CDL will be discussed in later sections of this paper.

2.2 Flight Control Technologies and Operations

Aircraft autonomy varies dramatically amount unmanned aircraft. At one extreme, aircraft operate entirely by remote control. If the pilot is taken out of the loop, the aircraft would crash. At the other extreme, the aircraft may be controlled by an autopilot from takeoff to landing. The pilot-in-command stays outside of the C2 loop, but may intervene in the case of an emergency to override the autopilot.

Almost all modern UAS are equipped with both manual remote control and autopilot technologies. Flexibility to operate either way make a UAS more marketable as it fits the need of wide range of customers. From our study, a generalization could be made that the control technology for which the UAS operates under is highly specific to the mission characteristics and the ability of the aircraft to operate RF beyond line-of-sight.

Table 1 Line-of-sight communication for a sample of surveyed unmanned aircraft

Aircraft	Manufacturer	LOS Communication	Characteristics
Predator	General Atomics Aeronautical Systems	C-Band	Wing span: 20.1 m Length: 10.9 m Payload: 385.5 kg Max. altitude: 15,240 m Max endurance: 30 h
Global Hawk	Northrop Grumman Integrated Systems	CDL (137 Mbps, 274 Mbps); UHF SATCOM	Wing span: 39.9 m Length: 14.6 m Payload: 1,360.7 kg Max. altitude: 18,288 m Max endurance: 36 h
ScanEagle	Insitu Group	900 MHz of spread spectrum frequency hopping; UHF command/telemetry	Wing span: 3.1 m Length: 1.2 m Payload: 6.0 kg Max. altitude: 4,998 m Max endurance: 20 h
Meridian	University of Kansas	72MHz Futaba radio 16 km; 2.4GHz microband radio	Wing span: 8 m Length: 5.1 m Payload: 54.4 kg Max. altitude: 4,572 m Max endurance: 9 h
Desert Hawk	Lockheed Martin	Military 15 km data link	Wing span: 1.4 m Length: 0.9 m Payload: 3.1 kg Max. altitude: 152 m Max endurance: 1 h
Dragon Eye	AeroVironment	Military 10 km data link @ 9,600 baud	Wing span: 1.2 m Length: 0.9 m Payload: 2.3 kg Max. altitude: 152 m Max endurance: 1 h
Manta B	Advance Ceramic Research	Military band/ISM band radio modem; 24–32 km radio	Wing span: 2.7 m Length: 1.9 m Payload: 6.8 kg Max. altitude: 4,876 m Max endurance: 6 h

For LOS operations, a low endurance UAS typically uses remote control for part if not all of the flight. For takeoff and landing, a remote pilot will control the aircraft. Once airborne, the pilot may decide to fly the aircraft manually for the entirety of the flight path or allow the autopilot to perform simple waypoint navigation along a flight path. Some common UAS operating in such manner are Manta B, Meridian, Raven A and B, Dragon Eye, Silver Fox, and Shadow.

On the contrary, the high endurance UAS typically use autopilot for all LOS flight operations. The entire flight plan is programmed into the autopilot through a user interface at the ground control station. Once the mission begins, the aircraft will autonomously takeoff and follow the predefined path. The pilot remains out of the C2 loop, but monitors the flight operations for unusual situations, and, if need be, interrupts the autopilot and takes the manual remote C2. The Predator, Marnier,

ScanEagle, Georanger, Global Hawk, and BAMS are the example of UAS typically using autopilot technologies.

2.3 Link-Loss Procedures

The loss of a data link must be addressed by a link-loss procedure. It is important that the aircraft always operates in a predictable manner. From the survey, it was revealed that the most common link-loss procedure is for the aircraft to fly to a predefined location. Once at the predefined location, the UAS can either loiter until the link is restored, it can autonomously land, or it can be remotely piloted via secondary data link [21, 25, 27, 30]. In this section, specific examples are discussed.

The BAT III UA's link loss procedure involves a simple return home functionality, where it turns to the last known location of the ground control station and flies directly toward it [27]. A backup remote control radio at 72 MHz is available onboard the aircraft. Once within sufficient range to the base, a remote pilot will control the aircraft to land.

NASA and Boeing PhantomWorks X-36 follows a similar method of returning to base and loitering [30]. Rather than simply return to the base directly, the aircraft follows a pre-defined return path. To guide the UA to the path, steering points also exist. This proved troublesome as the X-36 would choose the nearest steering point, which in some cases may be behind the aircraft.

Similar to the X-36 return procedure, researchers at NASA Dryden have worked on a path planning algorithm for return-to-base and loss link operations that ensures the UA stays within its authorized flight zone [21]. Prior to departure, a return flight path was defined for use in the event of a loss of the control data link. Onboard autonomy will detect the loss of the link, and then maneuver the aircraft toward the return flight path.

Officials at Fort Carson have drafted a document for *Unmanned Aerial Vehicle Flight Regulations*. The military base includes two potential flight areas, one is restricted airspace, and the other is non-restricted airspace requiring either a Certificate of Authorization (CoA) or Temporary Flight Restriction (TFR) from the FAA [29]. They defined their classes of unmanned aircraft as Tactical UAV (TUAV) for operation beyond visual line-of-sight or over 1,000 ft; and Small UAV (SUAV) for flight under 1,000 ft or within visual line-of-sight. For the restricted airspace, if a TUAV loses link, it returns to a predefined grid location and loiters at 8,000 ft. If the SUAV loses link in the restricted airspace, it returns to the center of mass of the restricted airspace and lands. In both cases, necessary military authorities are contacted. When operating under CoA or TFR, the procedures modified in that FAA or other civilian authorities will be notified. If in either case the aircraft is likely to leave its restricted airspace, the flight will be terminated by some undisclosed means.

Finally, as disclosed from the NTSB report on the US Border Patrol Predator-B crash, the Predator-B will follow some predefined course back to base to loiter until communications are restored. In this case, as with others, the loss link procedure has on occasions been initiated for reasons besides an actual loss of data link [25]. In this case, operators attempted to use the link-loss procedure to restore the aircraft to a known working state.

2.4 ATC Communication and Coordination

ATC communications is one of the key factors allowing safe flight operations in NAS. It is utmost important that a UAS remains in contact with ATC during its entire mission even if traveling long distances from the ground control station. The literature research revealed that most ATC communications occur via a VHF transceiver. Special phone numbers were also setup in some instances to provide immediate contact with ATC in the event of a loss of data link.

In case of low endurance UAS for LOS operations, the pilot at ground control station is directly linked with ATC via VHF link. Redundant systems are employed for this purpose; a land based phone line is one of the options.

Our survey revealed very few details related to specific aircraft operations. Currently, unless operating under restricted military airspace, a Special Airworthiness Certificate or Certificate of Authorization (CoA) was required. To obtain such permission, procedures for interaction with ATC must be coordinated prior to flight. New interim regulations discussed later in this paper spell out more specific requirements for future UAS.

2.4.1 ATC Link Loss Procedures

ATC communication link loss procedures are handled quite differently from C2 link loss procedure. The primary objective in case of ATC link loss is to re-establish voice communication between the ground control station and the ATC facility. So in case of loss of direct link with ATC for LOS operation, a land based phone line is the only option currently used. Some unmanend aircraft are also equipped with multiple VHF transceivers that could be used to establish a ground control-aircraft-ATC voice communication link. Most link loss occurs due to weather so it is very likely that the later option is also unavailable, as all VHF voice communications will be affected in the same vicinity.

3 Beyond RF Line-of-Sight C3 Technologies and Operations

The survey of beyond line-of-sight UAS covers primarily high endurance UAS, but a few medium endurance UAS that operate beyond line-of-sight. In the first category, Predator [7], Marnier [7], Global Hawk [24], and BAMS [24] are surveyed. For the second category, Meridian [10], ScanEagle [15], and Georanger [15] are studied. Table 2 lists some examples of BLOS C3 Technologies aboard UAS.

3.1 Data Links

BLOS C2 data links range from Ultra High Frequency (300 MHz) to Ku Band (15 GHz) [23]. Ku Band SATCOM data links are widely used for BLOS C2 system. It has a frequency range from 11.7–12.7 GHz for downlink and 14–14.5 for uplink [28]. Ku Band is used by a bulk of high endurance UAS like Global Hawk, BAMS, Predator and its derivatives. INMARSAT SATCOM data links are also used by high endurance UAS including BAMS, Marnier and Global Hawk. It has frequency range from 1,626.5–1,660.5 MHz for uplink and 1,525–1,559 MHz for downlink [14]. L Band Iridium SATCOM data links are used by smaller, low or medium endurance,

Table 2 Beyond line-of-sight communication for a sample of surveyed unmanned aircraft

Aircraft	Manufacturer	BLOS Communication	Characteristics
Predator	General Atomics Aeronautical Systems	Ku-Band SATCOM	Wing span: 20.1 m Length: 10.9 m Payload: 385.5 kg Max. altitude: 15,240 m Max endurance: 30 h
Global Hawk	Northrop Grumman Integrated Systems	Ku-Band SATCOM; Inmarsat	Wing span: 39.9 m Length: 14.6 m Payload: 1,360.7 kg Max. altitude: 18,288 m Max endurance: 36 h
ScanEagle	Insitu Group	Iridium	Wing span: 3.1 m Length: 1.2 m Payload: 6.0 kg Max. altitude: 4,998 m Max endurance: 20 h
Meridian	University of Kansas	Iridium A3LA-D Modem 2.4 Kbits/s 1,616–1,626.5 MHz	Wing span: 8 m Length: 5.1 m Payload: 54.4 kg Max. altitude: 4,572 m Max endurance: 9 h
Desert Hawk	Lockheed Martin	No BLOS Operations Disclosed	Wing span: 1.4 m Length: 0.9 m Payload: 3.1 kg Max. altitude: 152 m Max endurance: 1 h
Dragon Eye	AeroVironment	No BLOS Operations Disclosed	Wing span: 1.2 m Length: 0.9 m Payload: 2.3 kg Max. altitude: 152 m Max endurance: 1 h
Manta B	Advance Ceramic Research	No BLOS Operations Disclosed	Wing span: 2.7 m Length: 1.9 m Payload: 6.8 kg Max. altitude: 4,876 m Max endurance: 6 h

research UAS. It has a frequency range from 390 MHz–1.55 GHz [17]. Georanger, a medium endurance unmanned aircraft, and Meridian, a research unmanned aircraft, use Iridium modems as part of the avionics communication package.

Investigating satellite communication providers, low earth orbiting (LEO) and geosynchronous earth orbiting (GEO) satellites represent two extremes. LEO satellites operate at an altitude of 1500 km. GEO satellites operate at an altitude of 35000 km. In [26], a constellation of 80 LEO satellites was compared with a six satellite GEO constellation with equivalent coverage area using Ka Band. The LEO constellation outperformed the GEO constellation with reduced latency, lower path losses, and reduced launch cost. A LEO satellite constellation does have higher operational costs. As satellites pass overhead, service may be temporarily disrupted as the communication is automatically handed-off to the next satellite. Examples of widely used LEO constellations include Iridium [17] and Globalstar [9].

One potential data link under consideration for BLOS communication is the military's existing Common Data Link [8]. It is unknown whether such a system would be applicable for civilian use or not, but it appears that CDL links are used on many of the larger UAS operated by the military;e.g. Predator-B, Global Hawk, etc. While no documents explicitly support this claim, the aircrafts' previously cited data sheets show identical specifications without explicitly stating that it is a CDL link. In addition to CDL, there is also the Tactical CDL, which also includes additional security.

Two technologies exist for CDL links. The first uses an I-band satcom link. The second data link uses Ku Band at 14.5–15.38 GHz in order to increase available bandwidth [8].

3.2 Flight Control Technologies and Operations

Satellite-based communications as are discussed in the last section are the primary means of beyond line-of-sight command and control communication with unmanned aircraft. Latency is a key issue encountered with SATCOM data links. At such latencies, remote piloting becomes less feasible. Autopilots are therefore required for control of most UAS under beyond RF line-of-sight operations.

The high endurance UAS uses autopilot for all flight operations. Entire flight plan is programmed in the autopilot GUI. The pilot remains out of the C2 loop but monitors the flight operations for unusual situations. A prime example of such a mission was the Altair UAS that flew a restricted flight plan imaging wildfires in western states. Altair is a NASA owned UA used for research purposes; a CoA was provided for each flight based on FAA approval of flight plans submitted by NASA [5]. Unfortunately, specific details of autopilot design and operations for the majority of high-endurance UAS were not disclosed.

The medium endurance UAS has the opportunity of using either LOS or BLOS technologies during the course of its flight. The research unmanned aircraft Merdian from the University of Kansas provided key insight into the transitions between LOS and BLOS and how operations change. Meridian is integrated with an autopilot for command and control [10], which allowed both remote piloting and waypoint navigation. Currently, a WePilot [31] autopilot is utilized (W. Burns, Personal communications, University of Kansas, 2008). When within close proximity, a 2.4 GHz LOS modem may be used and the pilot will receive real-time telemetry such that he may visualize the aircraft and control it similar to Instrument Flight Approach. Once the aircraft has exceeded the 2.4 GHz radio's range, the aircraft transitions to its Iridium data link. While using the Iridium due to latency and bandwidth, operation is limited to high-priority aircraft telemetry and waypoint navigation (W. Burns, Personal communications, University of Kansas, 2008) [10].

3.3 Lost-Link Procedures

From the survey, it was found that link loss procedures for BLOS operation in either medium endurance or high endurance unmanned aircraft are nearly identical to LOS operations. When the avionics detect a loss of a data link, a preplanned control sequence is executed. The aircraft either returns to base or emergency rendezvous point, taking a pre-determined flight path, or it loiters within a pre-determined airspace.

Altair [5] flew in NAS for Western States Fire Imaging Mission with CoA from FAA. During one of its mission, the Altair had modem malfunction, resulting in BLOS Ku Band C2 link loss. As a result, the aircraft switched to C Band and flew to pre-determined air space until the modem returned to normal functioning and the Ku Band link was established [5].

3.4 ATC Communication and Coordination

Since high endurance UAS may operate across multiple ATC regions, the current paradigm for ATC communication is to utilize the UAS as a communication relay. The relay allow a ground operator to remain in constant contact with the ATC of the UAS's current airspace. As part of a UAS's Certificate of Authorization, it is also a standard procedure to coordinate with all ATC sites along the UAS's path [5].

As discusses previously, the primary objective in case of ATC link loss is to re-establish the communication data link between ground control station and the ATC facility. For BLOS operations this is only possible by carrying multiple VHF transceivers and having redundant voice communication systems on board. This is necessary because as the aircraft travels through various ATC regions it contacts local ATC facility and the ground control station is connected to ATC via the aircraft as a relay.

For the Altair link loss, the FAA and ATC were provided with detailed flight plans, making sure that the ATC knew aircraft's location. Additionally, the missions were planned meticulously with respect to ATC coordination, such that all potential ATC facilities are notified. The mode of notification was not explicitly disclosed [5].

Using the unmanned aircraft as a relay between the ground control station and the overseeing ATC facility is not without several issues. The handoff of the aircraft between facilities presents an issue. For manned aircraft, as it transitions from one ATC cell to another, the onboard pilot dials the VHF radio to the appropriate ATC channel as instructed through the handoff procedure. For several existing CoAs and aircraft, the aircraft perform a rapid assent to uncontrolled airspace and maintain this altitude for the duration of the flight. As a result, interaction along a flight path involving multiple ATC facilities is not common, and proper procedures to handle operations within controlled airspace has not been formally developed. For UAS to operate within ATC controlled airspace in the NAS beyond line-of-sight, further protocols must be established regarding the handling of the handoffs, and setting of the new frequencies of the aircraft's ground-to-ATC relay.

Another potential issue of using UAS as a relay is the spectrum availability to handle the additional 25 KHz voice channels needed to support each unmanned aircraft (S. Heppe, Personal communications, Insitu, 2008). A proposed alternative would be the inclusion of a ground-based phone network that connects ground stations to the ATC facility for which the UA is operating under. These issues must be studied further.

4 Security Issues of UAS C3 Technology and Operations

Data link spoofing, hijacking, and jamming are major security issue facing UAS C2 and ATC communications. UAS are different than conventional aircraft from the point of "immediate control" of the aircraft. Pilot in "immediate control" means in an adverse event, the pilot can fly without putting aircraft or other aircraft in the immediate vicinity at risk of collision. In case of UAS, there is a medium between pilot at the ground control station and aircraft which is not the case with conventional aircraft. This medium being the data link, it is easily susceptible to threats mentioned previously. A hacker can create false UAS signals, jam the data link or even hi-jack the data link and take the control of UA. This issue must be addressed while picking the appropriate data link for future UAS C2 and ATC communication, as data links are vital to the safety and seamless functioning of the UAS.

In order to make C2 and ATC communication foolproof, security features can be built into the system. For example, one approach is for the aircraft to acknowledge or echo all commands it receives. This will ensure the pilot-in-command that all commands sent are received and acknowledged (S. Heppe, Personal communications, Insitu, 2008). Such an approach will also notify the pilot in command if the aircraft receives commands from an unauthorized entity. The military uses secured data links like CDL and Link 16 [20] with built-in validating functions. No such permanent solution is available for civilian market and the area must be explored.

5 Relevant Existing Protocols, Standards, and Regulations

5.1 Standardization Agreement 4586

NATO Standardization Agreement (STANAG) 4586 [22] was ratified by the NATO countries as an interoperability standard for unmanned aircraft systems. Now in its second edition, it was made available in April 20, 2004. One of the goals of STANAG 4586 is to achieve the highest defined level of interoperability of unmanned aircraft from NATO member nations. Several levels of interoperability are defined, and throughout the document, the relationships to these interoperability levels are indicated. It must be noted that the standard covers UAS command and control (C2) as well as payload communications.

STANAG 4586 is divided into two annexes. The first annex provides a glossary to support the second annex. The second annex provides an overview of the communication architecture, which is supported by three appendices. Appendix B1 discusses the data link interface. Appendix B2 discusses the command and control interface. More specifically B2 covers the military architecture that connects the ground control station with the military command hierarchy. Finally, Appendix B3 discusses the human and computer interfaces (HCI). For this study, the Annex B introduction and Appendix B1 are relevant.

Figure 3 presents the reference architecture for STANAG 4586. The UA is comprised in the UAV Air Component. Our interests lie in the Air Vehicle Element (AVE) as payload is not included as part of this technology survey. The AVE communicates with the UAV Surface Component (or ground station) through some data link. This Surface Component interacts with any external C4I infrastructure, any

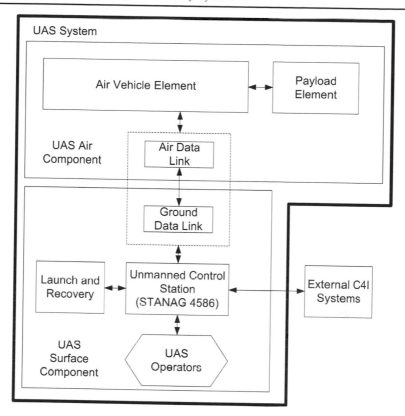

Fig. 3 STANAG 4586 system architecture [22]

ground operators controlling the aircraft, and the launch and recover system. The well defined interfaces between these components is presented as a block diagram in Fig. 4. Each interface is embodied in a system module that operates as a wrapper to support previously incompatible technologies.

For this investigation, the Vehicle System Module (VSM) defines the interface between the control station and the aircraft. Its Data Link Interface (DLI) is studied further as it defines the message set for command and control to the aircraft and telemetry from the aircraft. Figure 5 illustrates the interaction between the aircraft and ground station via STANAG 4586 interfaces.

STANAG 4586 has strong potential as a standard for UAS command and control in the National Airspace System. It has already been adopted by NATO member countries as the current interoperability standard for military UAS. Unlike JAUS [19], discussed in the next section, which was developed to cover all unmanned systems (ground, aerial, surface, and underwater), STANAG 4586 is specifically written to support UAS and has grained broader acceptance in the UAS industry than JAUS.

From the certification point-of-view, the adoption of an interoperability standard could dramatically simplify the certification process. The natural division of UAS systems by STANAG 4586 supports the ability to certify these components inde-

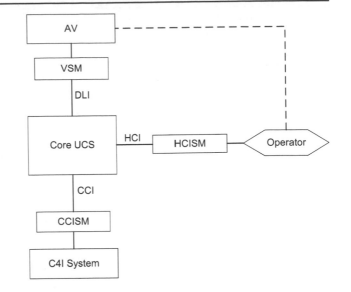

Fig. 4 STANAG 4586
components and interfaces
[22]

pendently. Unmanend aircraft can be certified as compliant independent of any ground station as its implementation of the VSM must support the data link interface. Likewise, the ground station may be certified by evaluating its interoperability with unmanned aircraft and operator control interfaces. Unfortunately, the FAA has not yet adopted modular certification.

There are some weaknesses of STANAG 4586. First, the standard includes messages that are specifically written to support communication with a UA's payload. Since payload is not incorporated into the certification process, these messages much be eliminated or modified. Fortunately, few messages incorporate both control and payload aspects.

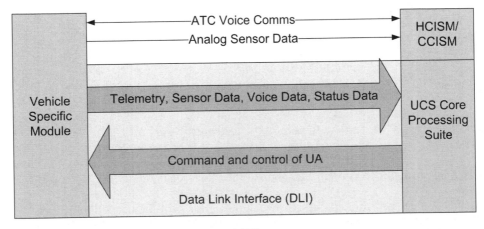

Fig. 5 STANAG 4586 ground station to aircraft [22]

5.2 Joint Architecture for Unmanned Systems

The Joint Architecture for Unmanned Systems (JAUS) [19] provides an alternate interoperability standard for unmanned aircraft. Unlike STANAG 4586, JAUS specifies interoperability of all unmanned systems and not just unmanned aircraft systems. Originally developed as a Department of Defense effort, JAUS is now being integrated into new standards by SAE through the AS-4 Technical Committee [16].

JAUS is an architecture defined for the research, development, and acquisition of unmanned systems. The documentation is divided into three volumes: Domain Model, Reference Architecture, and Document Control Plan [19]. The JAUS Working group [19] define in these volumes the architecture for representing unmanned systems, the communication protocols between these components, and the message set.

The technical constraints of JAUS are: platform independence, mission isolation, computer hardware independence, and technology independence [18]. The word System is used to define a logical grouping of subsystems. A subsystem is one or more unmanned system functions as a single localized entity within the framework of a system. A Node is defined as a distinct processing capability within a subsystem to control flow of message traffic. A component has a unique functionality capability for the unmanned system. Figure 6 presents a system topology using these terms.

Given the abstract definition of systems and subsystems and its applicability to unmanned systems other than aerial vehicles, JAUS does not seem to be sufficiently rigorous to meet the needs of regulators for defining a command and control protocol that may be certified for unmanned systems. As the architecture significantly changes toward new standards defined by AS-4, this concern may decrease, and warrants future consideration.

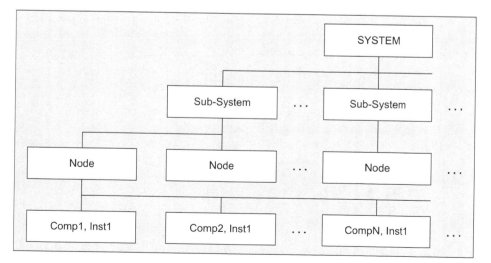

Fig. 6 Joint architecture for unmanned systems Topology [18]

5.3 International Civil Aviation Organization (ICAO) Annex 10

International Civil Aviation Organization is a United Nations agency responsible for adopting aviation standards and introducing as well as researching new practices for safety of international aviation. For the purpose of this technology survey, ICAO Annex 10 Aeronautical Telecommunications Volume I [12] and III [11] were researched. These standards apply to all aircraft and not only UAS. Volume III: Communication Systems discusses systems level requirements for various communication systems, namely aeronautical telecommunication networks, mobile satellite services, VHF air to ground digital data links, HF data links, etc., defining basic functionality, requirements and recommendations, and message protocols for developers.

5.4 FAA Interim Operational Approval Guidance 08-01

In March 2008, the FAA's Aviation Safety Unmanned Aircraft Program Office (AIR-160) with the cooperation of a number of other program offices at the FAA released interim guidelines for UAS operations in the NAS [6]. This document was produced to fill the need of some guidelines that may be shared between the FAA, UAS manufacturers, and UAS users until more permanent guidelines are put in place. The guidance may be used in order to assist in the distribution of Certificates of Authorization for public NAS users such as government agencies and Special Airworthiness Certificates for civilian NAS users. These regulations were meant to be flexible with periodic revisions.

The regulations do define some of the operational requirements of UAS relevant to Command, Control, and Communication. Since existing UAS detect, sense, and avoid (DSA) technologies are not safety certifiable at this time, a chase aircraft or ground observer is required during the operation of the UAS. The fact that some autonomy may exist for the unmanned aircraft is acknowledged and acceptable so long as a human-in-the-loop exists, or the system possesses a mechanism for immediate human intervention in its operation. Link-loss procedures must be defined for the aircraft such that it operates in a predictable manner. The guidance material does not suggest any particular link-loss procedure. If insufficient redundancy exists to maintain safe operation, a flight termination system may also be required. Such a system would be available to the pilot-in-command on the ground, or an observer in a chase aircraft.

Besides link loss procedures, control data links are not thoroughly discussed. The guidance material does specify requirements for communications with air traffic control. ATC communication must be maintained at all times. If the UAS is operating in instrument flight rules (IFR) under a pre-defined instrument flight path, it is required that the UAS possesses an onboard communication relay to connect its pilot-in-command (on the ground or in the chase aircraft) with the aircraft's local ATC. Lastly, under IFR, the UAS must possess at a minimum a mode C transponder (with a mode S transponder desired).

6 Conclusion and Future Work

From this research, an initial survey of technologies for command, control, and communication for unmanned systems was conducted. From this study, line-of-sight and beyond line-of-sight systems were compared. For each, the study addressed issues of communication technologies, command and control, lost link procedure, and ATC communications. This survey has assisted in gathering knowledge related to the technologies and issues associated with command, control, and communication.

The second phase of this project is a regulatory gap analysis. With a through understanding of current C3 technology, issues may now be enumerated with respect to FAA regulations, including, but not limited to, 14 CFR Part 21, 23, and 91 [1]. From these enumerated issues, the regulatory gaps will be analyzed. Regulatory gaps identify the areas of the existing regulations that are insufficient for the current state-of-the-art. Finally, the analysis will be shared with the FAA to hopefully assist in determining guidance material regarding airworthiness certification requirements for UAS users and manufacturers that wish for their aircraft to operate within the National Airspace System.

Acknowledgements This project was sponsored by the Federal Aviation Administration through the Air Transportation Center of Excellence for General Aviation, and was conducted by the members indicated. The Center of Excellence for General Aviation Research is comprised of the following universities: Embry-Riddle Aeronautical University, Florida A&M University, University of Alaska, University of North Dakota and Wichita State University. However, the Agency neither endorses nor rejects the findings of this research. The presentation of this information is made available in the interest of invoking technical community comment on the results and conclusions of the research.

Special thanks to Xiaogong Lee and Tong Vu of the FAA Research and Technology Development Branch—Airport and Aircraft Safety Team for their support.

References

1. 14 CFR: United States Code of federal regulations title 14 aeronautics and space. Online at: http://ecfr.gpoaccess.gov/cgi/t/text/text-idx?c=ecfr&tpl=/ecfrbrowse/Title14/14tab_02.tpl (2008)
2. AAI Corp.: Unmanned aircraft systems. Online at http://www.aaicorp.com/New/UAS/index.htm (2008)
3. Advance Ceramic Research: Unmanned vehicle systems. Online at http://www.acrtucson.com/UAV/index.htm (2008)
4. AeroVironment: Unmanned aircraft systems. AeroVironment Inc. Online at http://www.avinc.com/UAS_products.asp (2008)
5. Ambrosia, V.G., Cobleigh, B., Jennison, C., Wegener, S.: Recent experiences with operating UAS in the NAS. In: AIAA Infotech Aerospace 2007 Conference and Exhibit. Rohnert Park, California (2007)
6. FAA: Interim operational approval guidance 08-01: unmanned aircraft systems operations in the national arspace system. Technical report, Federal Aviation Administration. Aviation Safety Unmanned Aircraft Program Office AIR-160 (2008)
7. General Atomics: Aircraft platforms. General Atomics Aeronautical Systems Inc. Online at http://www.ga-asi.com/products/index.php (2008)
8. Global Security: Common data link. Online at http://www.globalsecurity.org/intell/systems/cdl.htm (2008)
9. Globalstar: Globalstar, Inc. — Worldwide satellite voice and data products and services for customers around the globe'. Online at http://www.globalstar.com (2008)
10. Hale, R.D., Donovan, W.R., Ewin, M., Siegele, K., Jager, R., Leong, E., Liu, W.B.: The meridian UAS: detailed design review. Technical Report TR-124, Center for Remote Sensing of Ice Sheets. The University of Kansas, Lawrence, Kansas (2007)
11. ICAO (ed.): Annex 10 communication systems, vol. 3, 1st edn. International Civil Aviation Organization (1995)
12. ICAO (ed.): Annex 10 radio navigation aides, vol. 1, 6th edn. International Civil Aviation Organization (2006)

13. ICAO: Fourth meeting of the AFI CNS/ATM implementation co-ordination sub-group. International Civil Aviation Organization. Online at http://www.icao.int/icao/en/ro/wacaf/apirg/afi_cnsatm_4/WP08_eng.pdf (2008)
14. INMARSAT: Aeronautical services. Online at http://www.inmarsat.com/Services/Aeronautical/default.aspx?language=EN&textonly=False (2008)
15. Insitu: Insitu unmanned aircraft systems. Online at http://www.insitu.com/uas (2008)
16. International, S.: Fact sheet SAE technical committee as-4 unmanned systems. Online at: http://www.sae.org/servlets/works/committeeResources.do?resourceID=47220 (2008)
17. Iridium: Aviation equipment. Online at http://www.iridium.com/products/product.htm (2008)
18. JAUS: Joint architecture for unmanned systems: reference architecture version 3-3. Technical Report, JAUS Working Group. Online at: http://www.jauswg.org/baseline/refarch.html (2007)
19. JAUS: Joint architecture for unmanned systems. JAUS Working Group. Online at: http://www.jauswg.org/ (2008)
20. Martin, L.: Tactical data links—MIDS/JTIDS link 16, and variable message format—VMF. Lockheed Martin UK—Integrated Systems & Solutions. Online at: http://www.lm-isgs.co.uk/defence/datalinks/link_16.htm (2008)
21. McMinn, J.D., Jackson, E.B.: Autoreturn function for a remotely piloted vehicle. In: AIAA Guidance, Navigation, and Control Conference and Exhibit. Monterey, California (2002)
22. NATO: NATO standarization agreement 4586: standard interfaces of UAV control system (UCS) for NATO UAV interoperability. North Atlantic Treaty Organization (2007)
23. Neale, M., Schultz, M.J.: Current and future unmanned aircraft system control & communications datalinks. In: AIAA Infotech Aerospace Conference and Exhibit. Rohnert Park, California (2007)
24. Northrop Grumman: Unmanned systems. Northrop Grumman Integrated Systems. Online at http://www.is.northropgrumman.com/systems/systems_ums.html (2008)
25. NTSB: NTSB incident CHI06MA121—full narrative. National Transportation Safety Board. Online at http://www.ntsb.gov/ntsb/brief2.asp?ev_id=20060509X00531&ntsbno=CHI06MA121&akey=1 (2008)
26. Peters, R.A., Farrell, M.: Comparison of LEO and GEO satellite systems to provide broadband services. In: 21st International Communications Satellite Systems Conference and Exhibit (2003)
27. Ro, K., Oh, J.-S., Dong, L.: Lessons learned: application of small UAV for urban highway traffic montoring. In: 45th AIAA Aerospace Sciences Meeting and Exhibit. Reno, Nevada (2007)
28. Tech-FAQ: What is C band?. Online at http://www.tech-faq.com/c-band.shtml (2008)
29. US Army: Unmanned aerial vehicle—flight regulations 95-23. Technical Report AR 95-23, The Army Headquarters. Fort Carson, Colorado (2005)
30. Walker, L.A.: Flight testing the X-36—the test pilot's perspective. Technical Report NASA Contractor Report no. 198058, NASA—Dryden Flight Research Center, Edwards, California (1997)
31. WeControl: WePilot 2000 technical brief. Online at: http://www.wecontrol.ch/pdf/wePilot2000Brief.pdf (2008)

Unmanned Aircraft Flights and Research at the United States Air Force Academy

Dean E. Bushey

Originally published in the Journal of Intelligent and Robotic Systems, Volume 54, Nos 1–3, 79–85.
© Springer Science + Business Media B.V. 2008

Abstract The United States Air Force Academy is actively involved in unmanned aircraft research across numerous departments involving many projects, aircraft, government agencies, and experimental programs. The importance of these research projects to the Academy, the faculty, the cadets, the Air Force, and to the defense of the nation cannot be understated. In an effort to be proactive in cooperating with recent concerns from the FAA about the growth and proliferation of UAS flights, the Air Force has implemented several new guidelines and requirements. Complying with these guidelines, directives, and regulations has been challenging to the researchers and research activities conducted at USAFA. Finding ways to incorporate these new guidelines effectively and efficiently is critical to research and participation in joint projects and exercises. This paper explores the nature of research at USAFA current restrictions imposed by the various regulations, the current process, short term solutions, and a long term vision for research into UAS at the Academy.

Keywords UAV · UAS · Unmanned aircraft · Research · Education · Air force · Flight · Airspace

1 Introduction

The United States Air Force Academy (USAFA) is actively engaged in several different areas of research involving Unmanned Aerial Systems (UAS), across many departments, supporting many different projects, agencies, and exercises. Several varieties of unmanned aircraft are flown at the academy, some of them commercially procured and modified for research, and some of them experimental one-of-a-kind

D. E. Bushey (✉)
Department of Computer Science, UAV Operations, US Air Force Academy, Colorado Springs, CO 80840, USA
e-mail: dean.bushey@usafa.edu

K. P. Valavanis et al. (eds.), *Unmanned Aircraft Systems*. DOI: 10.1007/978-1-4020-9137-7_6

aircraft. Recent changes to regulations; both FAA and military have necessitated a review of operating procedures of UAS flights at the Air Force Academy. This change has had, and will continue to have a major impact on research, education, development of these new technologies. The current process to approve these flights has arisen from the growing concern for the rapid growth in numbers and types of UAS and remote-controlled aircraft of various sizes airborne, and the safe flight of UAS by the US Military and other government agencies. These regulations, intended for larger scale operational flights, are being adapted for use with UAS, even micro controlled UAS systems used in several research projects.

In this paper we explore the nature of the research projects at USAFA, the goals, regulations affecting UAS operations, agencies involved. In addition, we examine both short term and long term solutions to these hurdles.

Lt Gen Regni, Superintendent of the USAF Academy recently highlighted the importance of UAS research at the Academy in an address to USAFA personnel [3]. The explosive growth in research, funding, scope and attention in recent years is expected to continue for the foreseeable future. Along with the research component of the Air Force Academy, one of the primary missions is producing future Air Force leaders that understand the nature of current technologies, uses, limitations, challenges, and possible avenues for growth and development.

2 Background

The United States Air Force Academy is an undergraduate engineering school designed to educate, train, and develop Cadets to become future leaders in the Air Force. Its primary focus is on developing cadets of character and leadership. There is no designated graduate research program at USAFA. However, one of the goals of USAFA is to encourage cadet and faculty research by developing and promoting leading-edge research into critical areas of interest to the US Air Force and the Department of Defense. Cadets are encouraged to explore many facets of UAS operations, including flight characteristics, cockpit design, human computer interaction, data transmissions, tactical use of UAS, persistent UAS, real-time video relay, automated visual and feature recognition, etc. A list of current projects is provided in attachment 1.

The Academy also supports and encourages faculty research and publication. Several researchers work with various other universities, agencies, and government departments on research and development into UAS and UAS related technologies. Part of this research helps to support many different organizations in the Department of Defense, Department of Homeland Security, intelligence agencies, Department of Immigration, and several other federal and state agencies.

2.1 Restricting Flight Operations at USAFA

In May, 2007 all UAV/UAS flights at the Academy were halted in response to guidance from the DoD and FAA requiring all UAS to have a Certificate of Authorization (COA) and to meet specific prescribed guidelines to fly in the national airspace system. On 24 Sep 2007, the Deputy Secretary of Defense signed a Memorandum of Agreement (MOA) with the FAA allowing the DoD to conduct

UAS flights without a COA provided that the flights abide by certain restrictions [1]. Current standard FAA regulations are two-tiered regarding UAS flights, one for civilian use and one for government use. Government use, which is broadly defined as public-owned aircraft operated in support of government operations, requires a Certificate of Authorization (COA) to fly in unrestricted airspace. This COA outlines airspace, altitude, and other operating restrictions and is generally granted per airframe as follows:

- Past experience has shown it takes approximately 6 months to receive a COA, but the FAA has agreed to a 60 day process for DoD requests (beginning 24 Sep 2007)
- The COA requires a detailed listing of aircraft performance, operational limitations and procedures, location of operations, and personnel authorized to operate the UAS under the COA.

   - The USAFA has submitted paperwork to the FAA for a blanket COA that would authorize all UAS flights within the AFA airspace. While this is under review Air Force Special Operations Command (AFSOC) is the lead agency in charge of UAS operations.
   - UAS operations with a valid COA require that the pilot and observer hold a valid class III medical certificate, and that the pilot holds a valid pilot certification issued by the FAA (Private Pilot's License, etc.)

- FAA/DoD Memorandum of Agreement, signed 24 Sep 2007 allows for flight of UAS over military reservations without a COA however,

   o Specific DoD guidance applies to all flights conducted over military reservations, and a COA is still required for flights outside of this airspace.
   o MOA delineates between micro UAS (smaller than 20 pounds) and other UASs

      - Micro UASs may operate within Class G airspace under 1,200 ft AGL over military bases provided the operations meet the following criteria:

         • UAS remains within clear visual range of the pilot, or a certified observer in ready contact with the pilot
         • UAS remains more than 5 miles from any civil use airport
         • DoD must publish a Notice to Airmen (NOTAM) no later than 24 h prior to the flight (blanket NOTAMs can be used if justified)
         • Pilots/Observers are qualified by the appropriate Military Department (AFSOC)

   o AFSOC has been designated lead command for developing small UAS guidance for qualification and operation for all USAF UAS.

      - Draft guidance released by AFSOC (AFSOC 11-2) is geared towards operational qualification of UAS pilots on Systems of Record intended for use in combat situations in close coordination with manned aircraft [2]. These requirements are being modified to meet the unique demands of USAFA researchers to meet due to time, manpower and money constraints

- AFRL has begun developing separate guidance for research and development UAS, however AFSOC is the designated lead and must approve any AFRL-developed guidance prior to implementation
- Current AFSOC guidance requires all UAS to obtain certification through a Special Operations Application Request (SOAR) process. Not originally designed for certifying micro UAS, this process is being adapted to accommodate experimental research UAS designed and used at USAFA.

2.2 AFSOC Guidance

AFSOC has published guidelines to follow for UAS operators seeking approval for National Airspace Operations. The published checklist states, "on September 24, 2007, FAA and DoD representatives signed a Memorandum of Agreement concerning the operation of DoD Unmanned Aircraft Systems (UAS) in the National Airspace System. Prior to approving such operations certain criteria must be met. The following checklist outlines the necessary requirements for the operation of DoD UAS in Class G Airspace ONLY. A "Yes" answer is required to the following statements in order to process your request." (AFSOC checklist for UAS operations)

1. This is a DoD or DOD-contracted UAS certified by one of the military departments as airworthy to operate in accordance with applicable DoD and Military Department standards
2. The UAS pilots, operators and observers are trained, certified and medically qualified by the appropriate Military Department to fly in Class G airspace.
3. The Unmanned Aircraft (UA) weighs 20 pounds or less.
4. The UA operations will be contained in Class G airspace, below 1200' AGL.
5. The Class G airspace is located over a military base, reservation or land protected by purchase, lease or other restrictions.
6. The UA will remain within clear visual range of the pilot, or certified observer in ready contact with the pilot, to ensure separation from other aircraft.
7. The UA operations will remain more than 5 miles from any civil use airport or heliport.
8. The applicant verifies that this operation has been thoroughly coordinated and approved with a government official within the unit and that the applicant has been appointed as the unit requesting authority.

2.3 Restrictions Imposed

The net results of these regulations and the military MOA is UAS research operations at the Academy must be revised to meet the new safety related guidelines. Currently applications for Special Operations Airworthiness Release (SOAR) approval on several UAS are under review. Guidance from AFSOC has been clarified to a certain extent, but hurdles remain.

2.3.1 Airworthiness Process

The SOAR process, as defined in the Airworthiness Circular dated 24 August 2007, spells out the criteria used to certify UAS as airworthy. The guidance for this process comes from several underlying regulations, including:

Air Force Policy Directive (AFPD) 62-6, *USAF Aircraft Airworthiness Certification*
Air Force Instruction 90-901, *Operational Risk Management*
Air Force Pamphlet 90-902, *Operational Risk Management (ORM) Guidelines and Tools*
Air Force Instruction 11-202V1, *Aircrew Training*
Air Force Instruction 11-202 V3, *General Flight Rules*
Airworthiness Certification Circular No. 4, *Certification Basis*
Airworthiness Certification Circular No. 5, *Risk Evaluation and Acceptance*
MIL-HDBK-514, *Operational Safety, Suitability, and Effectiveness (OSS&E) for the Aeronautical Enterprise*
MIL-HDBK-516, *Airworthiness Certification Criteria*
MIL-STD-882, *Standard Practice for System Safety*

- SOAR process through Aeronautical Systems Center (ASC) at Wright-Patterson AFB, certification:
  - The SOAR process is rapidly evolving to meet these new types of UAS, however the time line for some research projects may be too inflexible for such a review.
    - USAFA-led flight safety reviews, perhaps through the aeronautics department, might be faster, more flexible and still meet the spirit and intent of the FAA and AFSOC requirements
- Pilot Qualification
  - Rated Pilots, either military or FAA certified, may act as POC of flights.
  - USAFA may design student pilot designations as long as FAA ground school requirements have been met.
- UAS flights in other airspace
  - Ft Carson offers a restricted airspace that is available for use. This would preclude the need for COAs and SOAR process review. This is an interim solution as it is not feasible to schedule range time, commute 1.5 h each way to and from the range during the already too full academic day for cadets.
  - USAFA's own class D—initial consideration was given to extended the Academy's class D airspace to include the Cadet Area and proposed UAS flight airspace. From discussions with the FAA this may be very difficult to achieve, however it remains a topic of great interest.

3 The Way Ahead

It is essential to come up with a viable solution that meets the needs of the FAA and DoD to maintain safety requirements and to adhere to regulations, while at the same time allowing needed research and development to be conducted at USAFA. Possible solutions include:

– Academy Certification of micro UAS—certify an agency at USAFA, possibly the aerospace engineering department, as a certifier of safety of flight and COA for these micro UAS
– Class D airspace extension—as mentioned above, this is a possibility that would allow USAFA airfield managers to allow UAS flights with prior coordination. Modification of existing airspace designations is being pursued, but this process could take time, if approved at all.

3.1 Short Term Solutions

Working within the framework of the current process, several short term steps can be taken to streamline the approval process for UAS research at USAFA:

– Streamline and expediting of the current approval process if possible
– USAFA local approval authority for flight readiness review process for cadet capstone projects.
– Verbal Approval/Waiver Authority, pending through review of applications

3.2 Long Term Solutions

Many long term solutions have been proposed to ensure that this valuable research proceeds:

– Establishment of a USAFA UAS Research Center that would work closely with all agencies involved
– Establishment of a cadet UAS flight center that would allow all cadets the opportunity to experience UAS flight during their sophomore or junior years.

4 Summary

UAV/UAS research at the USAFA has becoming an increasing integral part of the cadet education. New, tighter restrictions and safety considerations have necessitated a review of operating procedures and temporarily halted certain research flights. Several people, agencies, and departments are working to solve the problems; high-level attention has been given to this issue; and the current process is being ironed out to facilitate research needs.

Samples of UAS Research by USAFA Faculty and Cadets. Revised 5 Apr 2008

Subject	Department	Faculty Point of Contact
Fly Eye	DFB with U of Wyoming	Dr. Mike Wilcox
Atmospheric Research (Requested by USAF Weather Service)	DFP	Dr. Matthew Mcharg
Viper Aircraft (*V*ersatile *I*ntegrated *P*latform for *E*xperimental *R*esearch)	DFAN	Lt Col Carl Hawkins
KC-135 Redesign–design, build and fly scale model	DFAN	Dr. Steve Brandt
Fighter Sized Target Study (design, build and fly scale model)	DFAN	Maj Jeremy Agte
SCARF (collect radio frequency data using software radios)	DFCS	Capt James Lotspeich
Situational Awareness Tool	DFCS	Dr. Steve Hadfield
Summer Space Program	DFCS	Capt James Lotspeich
Robust/Reliable UAV Platforms	DFAN/DFCS	Dr. Steve Brandt/ Capt James Lotspeich
Black Dart Exercise Support	DFCS and DFEM	Lt Col Bushey
Black Dart *Red Team* UAS Project	DFAN/Sys Engr	Lt Col James Greer
Improved UAV Batteries	DFC	Dr. John Wilkes
Multiple UAVs for Persistent ISR	DFEC	Dr. Daniel Pack
Intelligent Sensors for Persistent Tactical ISR	DFEC	Dr. Daniel Pack
Heterogeneous Active Sensor Network for Efficient Search and Detection of IED/EFP Associated Activities	DFEC	Dr. Daniel Pack
SECAF UAV Project Evaluate/Improve Predator Operator Control Stations	DFBL, DFCS, DFEC, Sys Engr	Lt Col David Bell
Micro Air Vehicles	DFEM	Dr. Dan Jensen

References

1. DOD–FAA Memorandum, 20070924 OSD 14887-07—DoD–FAA MoA UAS Operations in the NAS, 24 Sep (2007)
2. Air Force Special Operations Command Interim Guidance 11-2, UAV Operations, 1 Oct (2007)
3. Lt Gen Regni: Speech to USAFA personnel, 7 Mar (2008)

Real-Time Participant Feedback from the Symposium for Civilian Applications of Unmanned Aircraft Systems

Brian Argrow · Elizabeth Weatherhead · Eric W. Frew

Originally published in the Journal of Intelligent and Robotic Systems, Volume 54, Nos 1–3, 87–103.

Abstract The Symposium for Civilian Applications of Unmanned Aircraft Systems was held 1–3 October 2007 in Boulder, Colorado. The purpose of the meeting was to develop an integrated vision of future Unmanned Aircraft Systems with input from stakeholders in the government agencies, academia and industry. This paper discusses the motivation for the symposium, its organization, the outcome of focused presentations and discussions, and participant survey data from questions and statements that were collected in real time during the meeting. Some samples of these data are presented in graphical form and discussed. The complete set of survey data are included in the Appendix.

Keywords Unmannned aircraft system · UAS · Civil applications

1 Introduction

An investment of over 20 Billion dollars for the development of unmanned aircraft capabilities has resulted in U.S. leadership in the field and unprecedented capabilities that can now be successfully applied to civilian applications [11]. Unmanned Aircraft Systems (UAS) are highly developed based on decades of developments for modern

B. Argrow · E. W. Frew (✉)
Aerospace Engineering Sciences Department,
University of Colorado, Boulder, CO 80309, USA
e-mail: eric.frew@colorado.edu

B. Argrow
e-mail: brian.argrow@colorado.edu

E. Weatherhead
Cooperative Institute for Research in Environmental Sciences,
University of Colorado, Boulder, CO 80309, USA
e-mail: betsy.weatherhead@cires.colorado.edu

K. P. Valavanis et al. (eds.), *Unmanned Aircraft Systems*. DOI: 10.1007/978-1-4020-9137-7_7

aircraft. At this stage, a broad range of aircraft exist from small and lightweight ones that often hold cameras or remote sensors to large airplanes, with wingspans of over 130 feet that can hold advanced scientific equipment [11]. All have one quality that is invaluable to civilian applications: they can go where it might be dangerous to send a manned aircraft and they generally provide persistence beyond the capabilities of manned aircraft.

Unmanned aircraft (UA) are proving valuable in initial attempts at civilian applications. UA have flown far over the Arctic Ocean, flying low enough to take detailed measurements of the Arctic ice that could not be taken with manned aircraft [8]. UAS are successfully being used to support homeland security, assisting in border patrol missions [6], keeping our agents both safe and effective in their duties. Recent flights into hurricanes brought back important measurements that could not have been gathered any other way [1, 3, 4]. Other civilian applications for which UA have already been fielded include law enforcement [12], wildfire management [13], and pollutant studies [7]. These first attempts to use UAS for civilian applications have brought tremendous results while maintaining safety in the air and on the ground.

Atmospheric scientists and engineers at the University of Colorado convened the first ever community gathering of those interested in civilian applications of UAS [2]. Entitled *The Symposium for Civilian Applications of Unmanned Aircraft Systems (CAUAS)*, this meeting was held 1-3 October 2007 in Boulder, Colorado (Fig. 1). Leaders from industry, government and academia came together for three days to discuss successes, future needs, and common goals. Representatives from over thirty five industries, a dozen universities, and five agencies (NASA, NOAA, DHS, DOE and Federal Aviation Administration (FAA)) gathered to discuss, across disciplines, the priorities for the next ten years. The CAUAS Steering Committee determined that the next decade will be critical for the technologies, regulations, and civilian

Fig. 1 The symposium for CAUAS was held 1–3 October 2007 in Boulder, Colorado

applications that will drive the design and development of future UAS. Fundamental to the future success of these efforts is the development of the technology and infrastructure to allow civilian applications to proceed safely. The development of these capabilities will allow the U.S. to maintain its technological leadership while successfully addressing important societal needs to protect the homeland, improve our understanding of the environment, and to respond to disasters as they occur.

2 Symposium Structure

The Symposium was structured to answer the questions: 1) What are current US civilian capabilities, 2) Where does the civilian UAS community want to be in 10 years, and 3) How do we get there? Each day of the three-day symposium was structured to address one of the primary questions:

Day 1: Current US civilian capabilities: Successes and failures

1. Background
2. Public and Commercial Applications
3. Scientific Applications
4. Industry, FAA, agency Vision

Day 2: The Public Decade: Where do we want to be in ten years?

1. Scientific Applications
2. Public and Commercial Applications
3. Visions for the integration of UAS capabilities into scientific and public goals
4. Industry, FAA, agency Vision

Day 3: How do we get there? Steps in the Public Decade

1. Existing paradigms and new alternatives
2. Applications driven engineering challenges
3. Society and policy
4. Industry, FAA, and agency vision

On Day-1, Session 1 focused on how UAS technology has arrived at the current state of the art. The session started with presentations that summarized military applications and how the Department of Defense has been, and continues, to be the primary driver for UAS development and deployment. This was followed by a contrast of wartime and commercial applications of UAS, an overview of international UAS capabilities, and a view of the current state from the FAA perspective. Session 2 focused on recent public and commercial applications, both successes and failures, that included recent joint-agency (NASA, NOAA, USFS, FAA) missions to support firefighting efforts in the Western US, and DHS (Customs and Border Protection) operations along the US-Mexico border. Session 3 looked at current scientific applications ranging from sea-ice survey missions in the Arctic, to the measurement of various surface-to-atmosphere fluxes, to the integration of satellite-based remote sensing with UAS from observation platforms. Each day concluded with a Session 4 that provided a panel discussion where agency, academia, and industry representatives provided their assessment of what they heard during the previous three sessions.

The focus of Day-2 was the *Public Decade* where the term *public* follows the FAA definition that categorizes aircraft and aircraft operations as either *civil* or *public* [9]. Civil refers to commercial aircraft and operations. Public aircraft and operations are those sponsored by federal or state governments, that includes agencies and universities. In the first session, scientists and agency representatives presented their vision of science applications enabled by UAS technologies in 2018. In Session 2 presentations focused on visions for public and commercial applications such as search and rescue, border patrol, and communication networking. Session 3 focused on visions of integrated sensing systems that might be deployed for applications in 2018 such as in situ hurricane observations and fire monitoring. The day was concluded with a Session 4 panel discussion.

With the first two days addressing the state of the art and visions of the near future, the final Day-3 addressed the question of "How do we get there: What are the steps forward during the public decade?" This day was planned to focus on the technical engineering challenges, and also on the societal perception and regulations challenges. Presentations in Session-1 focused on alternatives to manned flight and regulatory constraints, who can use UAS, does their development makes sense from a business perspective, and a discussion of an UAS open-design paradigm. Session-2 presentations addressed applications-driven engineering challenges that include sensor development, payload integration, and command, control, communications, and computing challenges. The Session-3 presentations focused on the social and regulatory challenges with issues such as public perception and privacy, education of the public, and the challenge to US leadership with the global development and operations of UAS. Session 4 was conducted in the manner of the previous days, and the meeting was concluded by an address from the Symposium chairs.

3 Results of Real-Time Participant Feedback

A unique approach to understand the thoughts of the broad range of experts and measuring the collective opinions of the participants was employed for the Symposium. With over 150 people in attendance most days, hearing from each individual on a variety of subjects was not feasible. Clickers, donated from *i-clicker* (a Macmillan US Company, http://www.iclicker.com/), were used to poll the attendees on a variety of subjects. All the questions and responses are included in the Appendix and a few selected charts from the tabulated data are discussed below. Most often, statements, such as "UAS provide an irreplaceable resource for monitoring the environment" were presented and the audience could respond with one of five choices: Strongly Agree (SA), Agree (A), Neutral (N), Disagree (D), and Strongly Disagree (SD). For a small category of issues, the audience was asked to grade the community on specific issues, with the grades being A, B, C, D, and F, and in some cases the five responses corresponded to specific answers. The broad range of responses for the questions reflect the large diversity of the UAS civilian community who gathered in Boulder. The compiled results are broadly categorized as 1) participant demographics; 2) safety, regulations, and public perception; 3) applications, missions, and technologies; and 4) economics.

Figure 2 describes the demographics of the symposium participants and roughly approximates the targeted mix of participants sought by the CAUAS Steering

Committee invitations, with 36% from government agencies, 26% from academia, and 21% from industry. About half (45%) of the participants had more than 5 years of UAS experience.

Safety, regulatory issues, and public perception proved to be the primary focus of the discussions. There continues to be widespread misunderstanding of what is the current FAA policy for the operation of UAS in the United States National Airspace System (NAS). (Note that in March 2008, FAA published the most recent UAS policy revision [10]). Once this topic was raised and the current policies were discussed, it was clear that many continue to conduct research and commercial operations in violation of these policies and regulations. This revelation added to the urgency of the discussion for the CAUAS community to work more closely with FAA to abide by regulations and to pursue the legal requirements that the FAA representatives described for NAS operations. A specific recommendation was for

Fig. 2 Demographics of symposium; **a** participant demographics, and **b** number of years of experience working with UAS

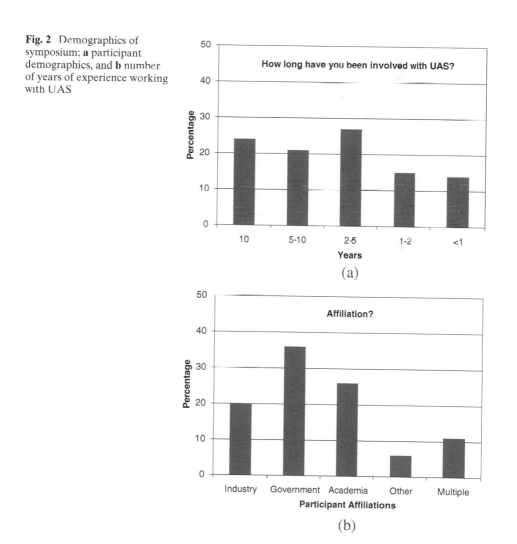

UAS researchers to seek locations with official restricted airspace to conduct legal flights with a minimum of FAA regulatory constraints.

Figure 3 shows some of the response to specific i-clicker questions. The response in Fig. 3a indicates the optimism of the participants with over 60% of the participants agreeing and strongly agreeing with the prediction that UAS operations in the NAS will become routine in the next 10 years. However, this is countered with pessimism that regulatory policies can keep pace with technological developments that might help with the integration of UAS into the NAS, with more than 80% of the participants expecting regulatory policies will not keep pace.

In discussions of the most likely public applications to be pursued over the next 10 years and those that will prove to be most important, disaster response was selected as the most likely application. Figure 4a shows the consensus of the group that there is a compelling case to use UAS for public applications. Figure 4b displays the

Fig. 3 Issues of safety, regulations, and public perception; **a** will operations in the NAS become routine, and **b** will NAS regulations keep pace with technological developments?

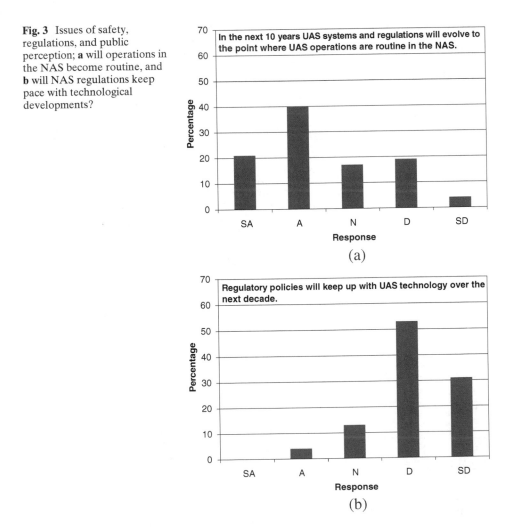

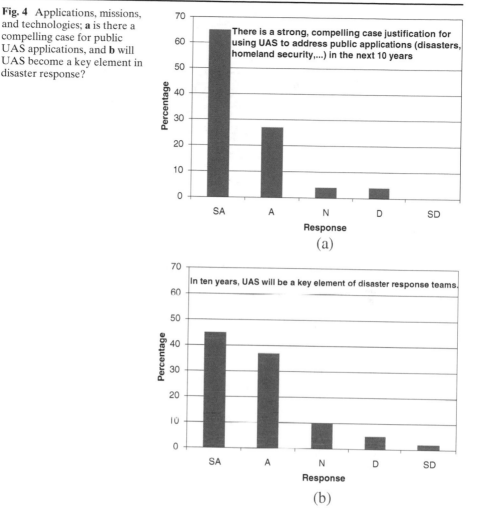

Fig. 4 Applications, missions, and technologies; **a** is there a compelling case for public UAS applications, and **b** will UAS become a key element in disaster response?

consensus that UAS will become an important and key element in disaster response teams during the next decade.

While Fig. 5a shows the general audience uncertainty that science applications will generate a significant UAS market, Fig. 5b shows strong agreement that market opportunities will be a primary driver if UAS are to be developed and applied in civilian applications. A participant pointed out that a current limitation is the lack of a civilian workforce trained to operate UAS. He suggested that this will begin to change with the return of U.S. service men and women who have operated UAS in theaters of conflict in Iraq and Afghanistan. Providing jobs for this workforce over the next decade might also have positive political payoffs to further enhance the economic future of UAS and their civilian applications.

Fig. 5 Applications, missions, and technologies; **a** is there a compelling case for public UAS applications, and **b** will UAS become a key element in disaster response?

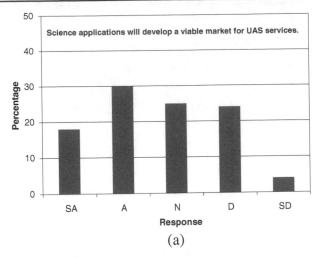

(a)

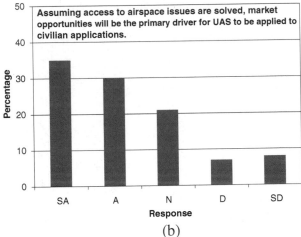

(b)

4 Outcome: UAS in the Public Decade

The CAUAS Steering Committee distilled the presentations and subsequent discussions into three civilian applications that are predicted to be of major importance during the Public Decade 2008–2018. These three applications are: 1) Disaster Response, 2) National Security, and 3) Climate Change. These applications were chosen, both because of their importance in which they were viewed in the participant feedback and because these are applications important to the agencies that will most likely have the financial support and the regulatory permission to carry them out.

For Disaster Response, UAS have already demonstrated their effectiveness for fire monitoring in six Western states. These preliminary flights helped fire-fighting efforts, directly saving costs, lives, and priority. UAS can also be used to survey disaster areas to provide immediate information for saving lives and property. Use of UAS after Hurricane Katrina could have expedited efforts to locate and rescue people stranded by floodwaters and dramatically helped in the disaster response.

UAS are becoming increasingly important for National Security applications focused upon protecting the homeland, helping Customs and Border Protection (CBP) streamline routine and event-driven operations along the countrys borders. Up to October 2007, CBP UAS helped personnel capture over 1,000 criminals crossing the Mexican border. Based on these successes, the Department of Homeland Security has significantly increased its investment in UAS.

UAS can fill an important gap not covered by existing systems for Climate Change observational applications. The observations would revolutionize the monitoring and scientific understanding of weather and climate. The Aerosonde, a small UAS manufactured and operated by AAI Corp., completed low-altitude penetration missions into both tropical cyclone Ophelia (2005) and Category-1 hurricane Noel (2007), providing real time observations of thermodynamic data and dangerous winds that in the future might be used in storm forecasting and storm track prediction [1, 3, 4]. UAS can also help measure carbon emissions providing necessary information for carbon trading.

The summary outcome of the meeting was condensed into an 8-page brochure entitled "CAUAS, Civilian Applications of Unmanned Aircraft Systems: Priorities for the Coming Decade [5]. An image of the cover page that captures the breadth of applications and technologies is shown in Fig. 6. The presentations

Fig. 6 Cover page for CAUAS brochure

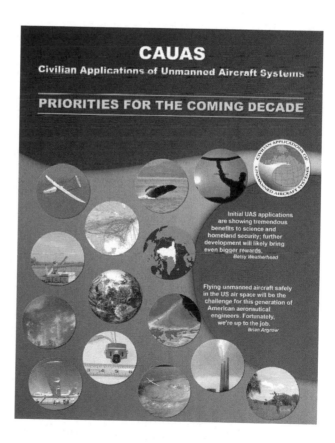

Fig. 7 Participants response to the statement: "The UAS community dedicated to civilian applications has a clear vision for how to use UAS in the coming decade." This statement was presented at the start of the meeting (*Monday*) and repeated at the close of the meeting (*Wednesday*)

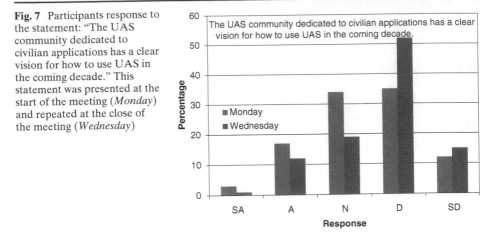

that formed the basis of this document can be found at the Symposium website http://cauas.colorado.edu.

To help gauge the impact of the meeting on the participants, the following i-clicker statement was asked on Monday, the first day of the symposium and again at the conclusion of the symposium on Wednesday: "The UAS community dedicated to civilian applications has a clear vision for how to use UAS in the coming decade". Figure 7 indicates that although a focus of the meeting was to discuss visions for the development and application of UAS technologies over the next 10 years, the participants indicated that they left with an even less clear vision of the future than before the meeting. There are at least two ways to interpret this result. First, it could mean that the presentation of a variety of visions for the future of UAS suggested the breadth of ideas of the possible. A second interpretation might be based on doubts that resulted from the clash of applications visions with the realities of regulatory requirements. In either case, the conclusion can be drawn that civilian community believes that there remains much uncertainty in the development of an integrated vision for the future of civilian applications of UAS.

5 Summary and Conclusions

CAUAS was the first symposium to bring together, in a single venue, the major stakeholders for the integration of UAS into the NAS for civilian applications, i.e., government agencies, industry, and academia. The meeting started with presentations that recounted the history of UAS and credited the current state-of-the-art to the development and maturation of military UAS. The remaining presentations focused on non-military, civilian applications and the need for conversion of military UAS for civilian applications, or for the development of UAS entirely for civilian applications. While technology has transfered between military and civilian uses, the view of the audience is that a separate profit-driven market must exist to support a UAS industry for civilian applications. Even if the number of such applications continues to increase, no substantial industry can currently exist unless UAS technologies can satisfy current FAA regulations developed for manned aircraft. This might soon be possible for larger UAS with UA that can carry equipment to emulate an onboard pilot and that can operate with the speed and performance of large manned aircraft. An alternative possibility is for new policies or regulations to be developed specifically for UAS integration of large UAS into the NAS. Either of these regulatory paths will continue to challenge or preclude the widespread integration of smaller UAS with performance requirements that require that they not operate like manned aircraft.

The CAUAS symposium was organized to focus on UAS applications for the *Public Decade*, 2008–2018; where "public" emphasizes the FAA distinction of civil and public aircraft. The CAUAS Steering Committee chose this as a reasonable time frame during which the government agencies might collaborate to take the lead and work with industry and academia to develop technologies, standards, and regulations to enable the integration of UAS into the NAS for civilian applications. If accomplished, this will expand opportunities and develop a significant market for the UAS industry. Of the applications presented and discussed, three rose to the level of a consistent organizing theme for to organize agency programs: Disaster Response, National Security, and Climate Change. These applications are summarized in the CAUAS summary document [2]. The purpose of this document is to lay out the case that can be made for civilian applications and to provide "policy relevant" data for those in positions to effect change.

The Symposium concluded with a general consensus that the gathered community had just begun to scratch the surface of the issues and obstacles to the integration of UAS into the NAS. Several recommended that CAUAS should reconvene for future symposia with specific emphasis on the civil and commercial applications and business cases, rather than the specific focus on public applications. At the time of this writing, the CAUAS organizers are considering a CAUAS-II meeting to be convened approximately two years after the original.

Acknowledgements The authors acknowledge the contributions of the CAUAS Steering Committee: Susan Avery, V. Ramanathan, Robin Murphy, Walt Oechel, Judy Curry, Doug Marshall, Sandy MacDonald, Steve Hipskind, Warren Wiscombe, Mike Kostelnik, Mike Francis, Raymond Kolibaba, Bob Curtin, Doug Davis, John Scull Walker, Steve Hottman, Mark Angier, and Rich Fagan. We also acknowledge the contributions of the CAUAS presenters whose presentations can be found at the Symposium website http://cauas.colorado.edu.

Appendix

Tables 1, 2, 3, and 4.

Table 1 CAUAS survey responses: demographics

	SA	A	N	D	SD
Affiliation: A) industry; B) government; C) academia; D) other; E) Multiple.	20	36	26	6	11
How long have you been involved with UAS? A) 10 years B) 5 to 10 years; C) 2 to 5 years D) 1–2 years; E) less than 1 year	24	21	27	15	14

K. P. Valavanis et al. (eds.), *Unmanned Aircraft Systems*

Table 2 CAUAS survey responses: safety, regulatory issues, and public perception

	SA	A	N	D	SD
In the next 10 years UAS systems and regulations will evolve to the point where UAS operations are routine in the NAS.	21	40	17	19	4
DOD-FAA MOU should be expanded to include other agencies.	37	32	19	5	6
Regulatory policies will keep up with UAS technology over the next decade.	0	4	13	53	31
Civilian applications of UAS should be consistent and compliant with FAA rules and regulations.	52	30	10	6	2
Safety standards applied to UAS are currently more stringent than manned aircraft.	20	31	16	24	9
UAS are still seen as a threat to the UAS community.	20	32	23	16	9
The clarification of UAS regulations have negatively impacted our UAS plans.	31	20	34	9	7
The clarification of UAS regulations have negatively impacted our UAS plans.	27	28	27	11	8
Your institution or agency would supply real cost data of manned flights, both direct and indirect, to assist the UAS industry help benchmark its costing.	12	32	33	12	10
The general aviation community is concerned that more regulations and equipment requirements will be levied on them as a result of UAS in the NAS.	22	34	29	15	1
UAS are seen as a threat to public safety.	13	22	31	22	11
UAS civilian applications: safety from UAS accidents.	22	22	42	11	2
UAS civilian applications: threat to National security.	15	24	35	18	9
UAS civilian applications: leadership.	1	3	17	47	31
UAS civilian applications: coordination.	0	1	24	43	32
UAS civilian applications: collaboration.	0	5	31	42	23

Table 2 (continued)

	SA	A	N	D	SD
UAS civilian applications: redundancies.	8	18	38	33	5
UAS civilian applications: education.	1	6	29	40	20
All UAS are aircraft and the FAA has the requirement per Title 49 (law) to set safety standards within the US designated airspace.	43	40	6	8	4
Unrealistic fears exist about UAS safety, ignoring excellent safety records of UAS (including Iraq).	9	31	22	32	6
The first UAS accident in the US that results in major losses of civilian lives will be a major setback to the entire civilian applications UAS community.	61	28	4	6	1
The UAS community dedicated to civilian applications has a clear vision for how to use UAS in the coming decade (Monday).	3	17	34	35	12
The UAS community dedicated to civilian applications has a clear vision for how to use UAS in the coming decade (Wednesday).	1	12	19	52	15

Table 3 CAUAS survey responses: applications, missions, and technologies

	SA	A	N	D	SD
UAS have shown critical value to homeland security and environmental threats.	49	38	9	3	1
Number 1 barrier to civil UAS systems: A) technology; B) safety; C) business case; D) spectrum; E) equity between airspace users.	5	62	15	6	12
The number of possible applications and possible platforms has not been fully explored yet for civilian benefits.	62	32	5	0	1

Table 3 (continued)

	SA	A	N	D	SD
Efforts to develop UAS capabilities for civilian applications are currently scattered across agencies, universities and companies with little coordination such as might be gained from a dedicated or central facility focused on UAS.	32	35	21	6	5
There is a strong, compelling justification for UAS to address important environmental science questions in the next ten years.	69	23	6	3	0
Optimal science capabilities in ten years will require preferential development of: A) sensors; B) platforms; C) infrastructure; D) air space access; all of the above.	5	1	3	49	41
Ability to fill gaps in capability will be the primary driver for government public UAS success (missions that satellites, manned aircraft and remote sensors cannot do).	43	33	8	13	3
In ten years, UAS will be a key element of disaster response teams.	45	37	10	5	2
There are NOAA environmental collection missions that can only be done with small UAS, which will drive small UAS routine operations by A) 2009; B) 2012; C) 2017; D) 2022; E) Not foreseeable.	22	41	24	7	6
There are high altitude environmental missions that can only be accomplished with the range, persistence and endurance of a HALE UAS. That requirement will drive	22	36	25	13	4

Table 3 (continued)

	SA	A	N	D	SD
NOAA development, integration and routine operation of HALE systems by: A) 2010; B) 2012; C) 2017; D) 2022; E) Not foreseeable.					
There is a strong, compelling justification for using UAS to address public applications (disasters, homeland security), in the next ten years.	65	27	4	4	0
Given appropriate sensor and platform development, science questions can be addressed in the next next ten years which cannot be addressed effectively any other way.	45	37	12	5	1
The UAS community will be a primary driver for air traffic management over the next decade.	1	22	16	40	21
The UAS community must engage the manned aircraft community e.g. AOPA as a consensus partner for UAS NAS integration over the next decade.	51	33	10	7	0
Alternative paths to expedite UAS integration into the NAS. Which is likely to be the most effective over the next 10 years? A) Use the power of the industry B) Bring the technology forward (with data) C) Provide draft policy and guidance to FAA D) Press the faa to certify systems for commercial use E) Government agencies support FAA.	13	35	24	7	21
What would be the most likely application to push UAS development forward? A) Agriculture; B) Water Resources; C) Climate and Weather; D) Law Enforcement; E) Emergencies and Disasters.	5	1	29	20	44

Table 4 CAUAS survey responses: economics

	SA	A	N	D	SD
The economic investment in environmental science in the next 10 years should rise to be commensurate with the problems being addressed.	60	21	14	4	1
Science applications will develop a viable market for UAS services.	18	30	25	24	4
In ten years, cost benefit will be the primary driver for successful commercial civil UAS applications. (profit)	42	35	14	7	2
In ten years, endurance/ persistence will be the most attractive attribute for UAVs to have commercial value (dull, dirty, dangerous).	15	34	24	24	3
Assuming access to airspace issues are solved, market opportunities will be the primary driver for UAS to be fully applied to civilian applications.	35	30	21	7	8
The economic investment in public applications (disasters, homeland security), of UAS in the next ten years should rise to be commensurate with the problems being addressed.	42	33	14	8	2

References

1. UAS: Final report: First-ever successful UAS mission into a tropical storm Ophelia—2005). http://uas.noaa.gov/projects/demos/aerosonde/Ophelia_final.html (2005)
2. CAUAS: Civilian applications of unmanned aircraft systems (CAUAS) homepage. http://cauas.colorado.edu/ (2007)
3. NASA: NASA and NOAA fly unmanned aircraft into Hurricane Noel. http://www.nasa.gov/centers/wallops/news/story105.html (2007)
4. NOAA: Pilotless aircraft flies toward eye of hurricane for first time. http://www.noaanews.noaa.gov/stories2007/20071105_pilotlessaircraft.html (2007)
5. Argrow, B., Weatherhead, E., Avery, S.: Civilian applications of unammned aircraft systems: priorities for the coming decade. Meeting Summary. http://cauas.colorado.edu/ (2007)
6. Axtman, K.: U.S. border patrol's newest tool: a drone. http://www.usatoday.com/tech/news/techinnovations/2005-12-06-uav-border-patrol_x.htm (2005)
7. Corrigan, C.E., Roberts, G., Ramana, M., Kim, D., Ramanathan, V.: Capturing vertical profiles of aerosols and black carbon over the indian ocean using autonomous unmanned aerial vehicles. Atmos. Chem. Phys. Discuss **7**, 11,429–11,463 (2007)

8. Curry, J.A., Maslanik, J., Holland, G., Pinto, J.: Applications of aerosondes in the arctic. Bull. Am. Meteorol. Soc. **85**(12), 1855–1861 (2004)
9. Federal Aviation Administration: Government Aircraft Operations. Federal Aviation Administration, Washington, DC (1995)
10. Federal Aviation Administration: Interim Approval Guidance 08–01: Unmanned Aircraft Systems Operations in the U. S. National Airspace System. Federal Aviation Administration, Washington, DC (2008)
11. Office of the Secretary of Defense. Unmanned Aircraft Systems Roadmap: 2005–2030. Office of the Secretary of Defense, Washington, DC (2005)
12. Sofge, E.: Houston cops test drone now in Iraq, operator says. http://www.popular mechanics.com/science/air_space/4234272.html (2008)
13. Zajkowski, T., Dunagan, S., Eilers, J.: Small UAS communications mission. In: Eleventh Biennial USDA Forest Service Remote Sensing Applications Conference, Salt Lake City, 24–28 April 2006

Computer Vision Onboard UAVs for Civilian Tasks

Pascual Campoy · Juan F. Correa · Ivan Mondragón ·
Carol Martínez · Miguel Olivares · Luis Mejías ·
Jorge Artieda

Originally published in the Journal of Intelligent and Robotic Systems, Volume 54, Nos 1–3, 105–135.
© Springer Science + Business Media B.V. 2008

Abstract Computer vision is much more than a technique to sense and recover environmental information from an UAV. It should play a main role regarding UAVs' functionality because of the big amount of information that can be extracted, its possible uses and applications, and its natural connection to human driven tasks, taking into account that vision is our main interface to world understanding. Our current research's focus lays on the development of techniques that allow UAVs to maneuver in spaces using visual information as their main input source. This task involves the creation of techniques that allow an UAV to maneuver towards features of interest whenever a GPS signal is not reliable or sufficient, e.g. when signal dropouts occur (which usually happens in urban areas, when flying through terrestrial urban canyons or when operating on remote planetary bodies), or when tracking or inspecting visual targets—including moving ones—without knowing their exact UMT coordinates. This paper also investigates visual servoing control techniques that use velocity and position of suitable image features to compute the references for flight control. This paper aims to give a global view of the main aspects related to the research field of computer vision for UAVs, clustered in four main active research lines: visual servoing and control, stereo-based visual navigation, image processing algorithms for detection and tracking, and visual SLAM. Finally, the results of applying these techniques in several applications are presented and discussed: this

P. Campoy (✉) · J. F. Correa · I. Mondragón · C. Martínez · M. Olivares · J. Artieda
Computer Vision Group, Universidad Politécnica Madrid,
Jose Gutierrez Abascal 2, 28006 Madrid, Spain
e-mail: pascual.campoy@upm.es

L. Mejías
Australian Research Centre for Aerospace Automation (ARCAA),
School of Engineering Systems, Queensland University of Technology,
GPO Box 2434, Brisbane 4000, Australia

K. P. Valavanis et al. (eds.), *Unmanned Aircraft Systems*. DOI: 10.1007/978-1-4020-9137-7_8   105

study will encompass power line inspection, mobile target tracking, stereo distance estimation, mapping and positioning.

Keywords UAV · Visual servoing · Image processing · Feature detection · Tracking · SLAM

1 Introduction

The vast infrastructure inspection industry frequently employs helicopter pilots and camera men who risk their lives in order to accomplish certain tasks, and taking into account that the way such tasks are done involves wasting large amounts of resources, the idea of developing an UAV—unmanned air vehicle—for such kind of tasks is certainly appealing and has become feasible nowadays. On the other hand, infrastructures such as oil pipelines, power lines or roads are usually imaged by helicopter pilots in order to monitor their performance or to detect faults, among other things. In contrast with those methods, UAVs appear as a cheap and suitable alternative in this field, given their flight capabilities and the possibility to integrate vision systems to enable them to perform otherwise human driven tasks or autonomous guiding and imaging.

Currently, some applications have been developed, among which we can find Valavanis' works on traffic monitoring [1], path planning for multiple UAV cooperation [2], and fire detection [3]. On the other hand, Ollero [4] has also made some works with multi-UAVs. There are, too, some other works with mini-UAVs and vision-based obstacle avoidance made by Oh [5] or by Serres [6]. Moreover, Piegl and Valanavis in [7] summarized the current status and future perspectives of the aforementioned vehicles. Applications where an UAV would manipulate its environment by picking and placing objects or by probing soil, among other things, can also be imagined and feasible in the future. In fact, there are plans to use rotorcraft for the exploration of planets like Mars [8, 9].

Additionally, aerial robotics might be a key research field in the future, providing small and medium sized UAVs as a cheap way of executing inspection functions, potentially revolutionizing the economics of this industry as a consequence. The goal of this research is to provide UAVs with the necessary technology to be visually guided by the extracted visual information. In this context, visual servoing techniques are applied in order to control the position of an UAV using the location of features in the image plane. Another alternative being explored is focused in the on-line reconstruction of the trajectory in the 3D space of moving targets (basically planes) to control the UAV's position [10].

Vision-based control has become interesting because machine vision directly detects a tracking error related to the target rather than indicating it via a coordinate system fixed to the earth. In order to achieve the aforemention detection, GPS is used to guide the UAV to the vicinity of the structure and line it up. Then, selected or extracted features in the image plane are tracked. Once features are detected and tracked, the system uses the image location of these features to generate image-based velocity references to the flight control.

In the following section briefly describe the different components that are needed to have an UAV ready to flight, and to test it for different applications. Section 3, explains with details the different approaches to extract useful information to achieve visual servoing in the image plane based on features and on appearance. Some improvements in 3D motion reconstruction are also pointed out. Section 4 describes visual control schemes employed to aim visual servoing, and the particular configuration of the control system assigned to close the visual control loop. Section 5 deals with the stereo configuration and theory to make motion and height estimation based on two views of a scene. Next, in Section 6 the simultaneous localization and Mapping problem based on visual information is addressed, with particular emphasis on images taken from an UAV. Section 7 shows experimental results of different applications, and Section 8, finally, deals with conclusions and future work.

2 System Overview

Several components are necessary to complete an operational platform equipped with a visual system to control UAVs. It is a multidisciplinary effort that encloses different disciplines like system modeling and control, data communication, trajectory planning, image processing, hardware architecture, software engineering, and some others. All this knowledge is traduced into an interconnected architecture of functional blocks. The Computer Vision Group at UPM has three fully operational platforms at its disposal, whereas two of them are gas powered Industrial Twim 52 c.c helicopters producing about 8 hp, which are equipped with an AFCS helicopter flight controller, a guidance system, a Mini Itx onboard vision computer, and an onboard 150 W generator. These helicopters are used for outdoors applications, as shown in Fig. 1, where one of the powered gas platforms performs an experimental autonomous flight. The third platform is a Rotomotion SR20 UAV with an electric motor of 1,300 W, 8A. It also has a Nano Itx onboard vision computer and WiFi ethernet for telemetry data. It is used on indoors and outdoors applications. In this section, a description of the main modules, their structure and some basic

Fig. 1 Aerial platform COLIBRI while is performing an experimental detection and tracking of external visual references

functionalities is provided. In general terms, the whole system can be divided into two components:

1. An onboard subsystem composed by:
    -   Vision computer with the image processing algorithms and related image treatment programs.
    -   Flight computer with Flight control software.
    -   Cameras.
    -   Communication interface with flight control and with ground subsystem.

2. A ground subsystem:
    -   Ground computer for interaction with the onboard subsystem, and data analysis.
    -   Communication interface.
    -   Data storage.

Those components' division can be reorganized into subsystems, which are described below.

2.1 Flight Control Subsystem

Most complex physical systems' dynamics are nonlinear. Therefore, it is important to understand under which circumstances a linear modeling and control design will be adequate to address control challenges. In order to obtain a linear dynamic model, the hover state can be used as a point of work to approximate the helicopter dynamics by linear equations of motion. Using this approximation, linearization around this state gives a wide enough range of linearity to be useful for controlling purposes.

The control system is based on single-input single-output (SISO) proportional-integral-derivative (PID) feedback loops. Such a system has been tested to provide basic sufficient performance to accomplish position and velocity tracking near hover flight [11–13]. The advantage of this simple feedback architecture is that it can be implemented without a model of the vehicle dynamics (just kinematic), and all feedback gains can be turned on empirically in flight. The performance of this type of control reaches its limits when it is necessary to execute fast and extreme maneuvers. For a complete description of the control architecture, refer to [14–17].

The control system needs to be communicated with external processes (Fig. 2) in order to obtain references to close external loops (e.g. vision module, Kalman filter for state estimation, and trajectory planning). The communication is made through a high level layer that routes the messages to the specific process. The next subsection introduces the communication interface in detail.

2.2 Communication Interface

A client-server architecture has been implemented based on TCP/UDP messages, allowing embedded applications running on the computer onboard the autonomous helicopter to exchange data between them and with the processes running on the ground station. The exchange is made through a high level layer which routes the messages to the specific process. Switching and routing a message depends on the type of information received. For example, the layer can switch between position

Fig. 2 Control system
interacting with external
processes. Communication is
made through a high level
layer using specific messages
routed for each process

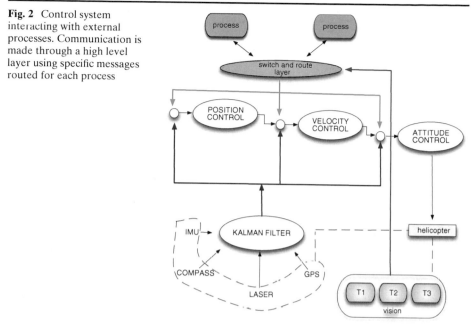

and velocity control depending on the messages received from an external process.
The mechanism used for this purpose consists in defining data structures containing a
field that uniquely identifies the type of information and the destination of a message.
Some of the messages defined for flight control are: velocity control, position control,
heading, attitude and helicopter state.

Figure 3, shows a case in which two processes are communicating through the
switching layer. One process is sending commands to the flight control (red line),
while the other one (blue line) is communicating with another process.

Fig. 3 Switching Layer.
TCP/UDP messages are used
to exchange data between
flight controller and other
process. Exchange is driven by
a high level layer which routes
the data to the specific process

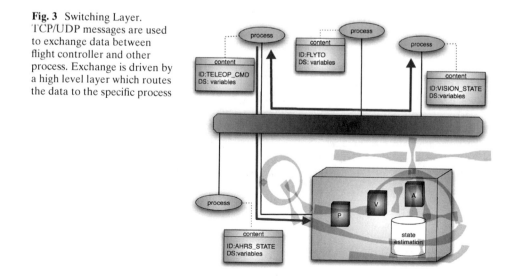

2.3 Visual Subsystem

The visual subsystem is a compound of a servo controlled Pan Tilt platform, an on-board computer and a variety of cameras and visual sensors, including analog/digital cameras (Firewire, USB, IP-LAN and other digital connections), with the capability of using configurations based on single, stereo cameras, arrays of synchronous multiple sensor heads and many other options. Additionally, the system allows the use of Gimbals' platforms and other kinds of sensors like IF/UV spectrum cameras or Range Finders. Communication is based on a long-range wireless interface which is used to send images for ground visualization of the onboard view and for visual algorithm supervision. Applications and approaches designed to perform visual tasks encompass optical flow, Hough transform, camera calibration, stereo vision to corner detection, visual servoing control implementation and Kalman filtering, among others.

Scene information obtained from image processing and analysis provides data related to the camera's coordinate system. This information is useful for purposes of automatic camera control, but not for the attitude and position control of the UAV. This issue is solved by fixating and aligning the camera's frame reference with the vehicle body-frame. Next section enumerates some basic algorithms of visual information extracted for controlling purposes.

3 Visual Tracking

The main interest of the computer vision group at UPM is to incorporate vision systems in UAVs in order to increase their navigation capabilities. Most of this effort is based on image processing algorithms and tracking techniques that have been implemented in UAVs and will be described below.

3.1 Image Processing

Image processing is used to find characteristics in the image that can be used to recognize an object or points of interest. This relevant information extracted from the image (called features) ranges from simple structures, such as points or edges, to more complex structures, such as objects. Such features will be used as reference for the visual flight control.

Most of the features used as reference are interest points, which are points in an image that have a well-defined position, can be robustly detected, and are usually found in any kind of images. Some of these points are corners formed by the intersection of two edges, and some others are points in the image whose context has rich information based on the intensity of the pixels. A detector used for this purpose is the Harris Corner detector [18]. It extracts a lot of corners very quickly based on the magnitude of the eigenvalues of the autocorrelation matrix. However, it is not enough to use this measure in order to guarantee the robustness of the corner, since the purpose of the features' extraction is to track them along an image sequence. This means that good features to track have to be selected in order to ensure the stability of the tracking process. The robustness of a corner extracted with the Harris detector can be measured by changing the size of the detection window, which is increased to test the stability of the position of the extracted corners. A measure of this variation is then calculated based on a maximum difference criteria. Besides, the magnitude of the eigenvalues is used to only keep features with eigenvalues higher than a minimum value. Combination of such criteria leads to the selection of the good features to track.

Another widely used algorithm is the SIFT (scale invariant feature transform) detector [19] of interest points, which are called keypoints in the SIFT framework. This detector was developed with the intention to use it for object recognition. Because of this, it extracts keypoints invariant to scale and rotation using the gaussian difference of the images in different scales to ensure invariance to scale. To achieve invariance to rotation, one or more orientations based on local image gradient directions are assigned to each keypoint. The result of all this process is a descriptor associated to the keypoint, which provides an efficient tool to represent an interest point, allowing an easy matching against a database of keypoints. The calculation of these features has a considerable computational cost, which can be assumed because of the robustness of the keypoint and the accuracy obtained when matching these features. However, the use of these features depends on the nature of the task: whether it needs to be done fast or accurate.

The use of other kind of features, such as edges, is another technique that can be applied on semi-structured environments. Since human constructions and objects are based on basic geometrical figures, the Hough transform [20] becomes a powerful technique to find them in the image. The simplest case of the algorithm is to find straight lines in an image that can be described with the equation $y = mx + b$. The main idea of the Hough transform is to consider the characteristics of the straight line not as image points x or y, but in terms of its parameters m and b. The procedure has more steps to re-parameterize into a space based on an angle and a distance, but what is important is that if a set of points form a straight line, they will produce sinusoids which cross at the parameters of that line. Thus, the problem of detecting collinear points can be converted to the problem of finding concurrent curves. To apply this concept just to points that might be on a line, some pre-processing algorithms are used to find edge features, such as the Canny edge detector [21] or the ones based on derivatives of the images obtained by a convolution of image intensities and a mask (Sobel [22], Prewitt). These methods have been used in order to find power lines and isolators in an inspection application [23].

3.2 Feature Tracking

The problem of tracking features can be solved with different approaches. The most popular algorithm to track features like corner features or interest points in consecutive images is the Lukas–Kanade algorithm [24]. It works under two premises: first, the intensity constancy in the vicinity of each pixel considered as a feature; secondly, the change in the position of the features between two consecutive frames must be minimum, so that the features are close enough to each other. Given these conditions to ensure the performance of the algorithm, it can be expressed in the following form: if we have a feature position $p_i = (x, y)$ in the image I_k, the objective of the tracker is to find the position of the same feature in the image I_{k+1} that fits the expression $p'_i = (x, y) + t$, where $t = (t_x, t_y)$. The t vector is known as the optical flow, and it is defined as the visual velocity that minimizes the residual function $e(t)$ defined as:

$$e(t) = \sum^{W}(I_k(p_i) - I_{k+1}(p_i + t))^2 w(W) \tag{1}$$

where $w(x)$ is a function to assign different weights to comparison window W. This equation can be solved for each tracked feature, but since it is expected that all features on physical objects move solidary, summation can be done over all features. The problem can be reformulated to make it possible to be solved in relation to all features in the form of a least squares' problem, having a closed form solution. In Section 3.3 more details are given. Whenever features are tracked from one frame to another in the image, the measure of the position is affected by noise. Hence, a Kalman filter can be used to reduce noise and to have a more smooth change in the position estimation of the features. This method is also desirable because it provides an estimation of the velocity of the pixel that is used as a reference to the velocity flight control of the UAV.

Another way to track features is based on the rich information given by the SIFT descriptor. The object is matched along the image sequence comparing the model template (the image from which the database of features is created) and the SIFT descriptor of the current image, using the nearest neighbor method. Given the high dimensionality of the keypoint descriptor (128), its matching performance is improved using the Kd-tree search algorithm with the Best Bin First search modification proposed by Lowe [25]. The advantage of this method lies in the robustness of the matching using the descriptor, and in the fact that this match does not depend on the relative position of the template and the current image. Once the matching is performed, a perspective transformation is calculated using the matched Keypoints, comparing the original template with the current image. Then, the RANSAC algorithm [26] is applied to obtain the best possible transformation, taking into consideration bad correspondences. This transformation includes the parameters for translation, rotation and scaling of the interest object, and is defined in Eqs. 2 and 3.

$$X_k = HX_0 \tag{2}$$

$$\begin{pmatrix} x_k \\ y_k \\ \lambda \end{pmatrix} = \begin{pmatrix} a & b & c \\ d & e & f \\ g & h & 1 \end{pmatrix} \begin{pmatrix} x_0 \\ y_0 \\ 1 \end{pmatrix} \tag{3}$$

where $(x_k, y_k, \lambda)^T$ is the homogeneous position of the matched keypoint against $(x_0, y_0, 1)^t$ position of the feature in the template image, and H is the homography transformation that relates the two features. Considering that every pair of matched keypoints gives us two equations, we need a minimum of four pairs of correctly matched keypoints to solve the system. Keeping in mind that not every match may be correct, the way to reject the outliers is to use the RANSAC algorithm to robustly estimate the transformation H. RANSAC achieves its goal by iteratively selecting a random subset of the original data points, testing it to obtain the model and then evaluating the model consensus, which is the total number of original data points that best fit the model. This procedure is then repeated a fixed number of times, each time producing either a model which is rejected because too few points are

Fig. 4 Experiments with planar objects in order to recover the full pose of the tracked object using SIFT. In the sub-figure **a** a template is chosen from the initial frame. In **b** the SIFT database is generated using the extracted keypoints. In **c** points are searched in a region twice the size of the template in the next image using the previous position as initial guess. **d** Subfigure shows the matching achieved by the tracking algorithm

classified as inliers or a model that better represents the transformation. If the total trials are reached, a good solution can not be obtained. This situation enforces the correspondences between points from one frame to another. Once a transformation is obtained, the pose of the tracked plane can be recovered using the information in the homography. Figure 4 shows an implementation of this method.

3.3 Appearance Based Tracking

Tracking based on appearance does not use features. On the other hand, it uses a patch of pixels that corresponds to the object that wants to be tracked. The method to track this patch of pixels is the same L–K algorithm. This patch is related to the next frame by a warping function that can be the optical flow or another model of motion. The problem can be formulated in this way: lets define X as the set of points that forms the template image $T(\mathbf{x})$, where $\mathbf{x} = (x, y)^T$ is a column vector with the coordinates in the image plane of the given pixel. The goal of the algorithm is to align the template $T(\mathbf{x})$ with the input image $I(\mathbf{x})$. Because $T(\mathbf{x})$ is a sub-image of $I(\mathbf{x})$, the algorithm will find the set of parameters $\mu = (\mu_1, \mu_2, \dots \mu_n)$ for motion model function $W(\mathbf{x}; \mu)$, also called the warping function. The objective function of the algorithm to be minimized in order to align the template and the actual image is Eq. 4

$$\sum_{\forall \mathbf{x} \in X} (I(W(\mathbf{x}; \mu) - T(\mathbf{x}))^2 \qquad (4)$$

Since the minimization process has to be made with respect to μ, and there is no lineal relation between the pixel position and its intensity value, the Lukas–Kanade algorithm assumes a known initial value for the parameters μ and finds increments of the parameters $\delta\mu$. Hence, the expression to be minimized is:

$$\sum_{\forall \mathbf{x} \in X} (I(W(\mathbf{x}; \mu + \delta\mu) - T(\mathbf{x}))^2 \qquad (5)$$

and the parameter actualization in every iteration is $\mu = \mu + \delta\mu$. In order to solve Eq. 5 efficiently, the objective function is linearized using a Taylor Series expansion employing only the first order terms. The parameter to be minimized is $\delta\mu$. Afterwards, the function to be minimized looks like Eq. 6 and can be solved like a "least squares problem" with Eq. 7.

$$\sum_{\forall \mathbf{x} \in X} \left(I(W(\mathbf{x}; \mu) + \nabla I \frac{\partial W}{\partial \mu} \delta\mu - T(\mathbf{x}) \right)^2 \tag{6}$$

$$\delta\mu = H^{-1} \sum_{\forall \mathbf{x} \in X} \left(\nabla I \frac{\partial W}{\partial \mu} \right)^T (T(\mathbf{x}) - I(W(\mathbf{x}; \mu))) \tag{7}$$

where H is the Hessian Matrix approximation,

$$H = \sum_{\forall \mathbf{x} \in X} \left(\nabla I \frac{\partial W}{\partial \mu} \right)^T \left(\nabla I \frac{\partial W}{\partial \mu} \right) \tag{8}$$

More details about this formulation can be found in [10] and [27], where some modifications are introduced in order to make the minimization process more efficient, by inverting the roles of the template and changing the parameter update rule from an additive form to a compositional function. This is the so called ICA (Inverse Compositional Algorithm), first proposed in [27]. These modifications where introduced to avoid the cost of computing the gradient of the images, the Jacobian of the Warping function in every step and the inversion of the Hessian Matrix that assumes the most computational cost of the algorithm.

Besides the performance improvements that can be done to the algorithm, it is important to explore the possible motion models that can be applied to warp the patch of tracked pixels into the $T(\mathbf{x})$ space, because this defines the degrees of freedom of the tracking and constrains the possibility to correctly follow the region of interest. Table 1 summarizes some of the warping functions used and the degrees of freedom. Less degrees of freedom make the minimization process more stable and accurate, but less information can be extracted from the motion of the object. If a perspective transformation is applied as the warping function, and if the selected patch corresponds to a plane in the world, then 3D pose of the plane can be

Table 1 Warping functions summary

Name	Rule	D.O.F
Optical flow	$(x, y) + (t_x, t_y)$	2
Scale+translation	$(1 + s)((x, y) + (t_x, t_y))$	3
Scale+rotation+ translation	$(1 + s)(R_{2x2}(x, y)^T + (t_x, t_y)^T)$	4
Affine	$\begin{pmatrix} 1 + \mu_1 & \mu_3 & \mu_5 \\ \mu_2 & 1 + \mu_4 & \mu_6 \end{pmatrix}$	6
Perspective	$\begin{pmatrix} \mu_1 & \mu_2 & \mu_3 \\ \mu_2 & \mu_5 & \mu_6 \\ \mu_7 & \mu_8 & 1 \end{pmatrix}$	8

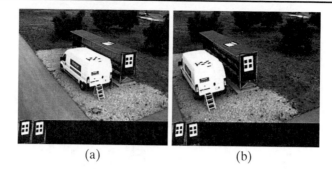

(a)   (b)

Fig. 5 Experiments using appearance based tracking were conducted to track a template in the scene. **a** Is the initial frame of the image sequence. Image region is manually selected and tracked along image sequence, using a scale + translation model (see Table 1). **b** Shows the tracked template 50 frames later from image (**a**). *Sub-images in the bottom of each figure* represent the initial template selected and the warped patch transformed into the template coordinate system

reconstructed from the obtained parameters. Figure 5 shows some tests carried out using a translation+scale motion model.

4 Visual Flight Control

4.1 Control Scheme

The flight control system is composed of three control loops arranged in a cascade formation, allowing it to perform tasks in different levels depending on the workspace of the task. The first control loop is in charge of the attitude of the helicopter. It interacts directly over the servomotors that define the four basic variables: cyclic/collective pitch of the principal rotor, cyclic/collective pitch of the tale rotor, longitudinal cyclic pitch, and the latitudinal cyclic pitch. The kinematic and dynamic models of the helicopter relate those variables with the six degrees of motion that this kind of vehicle can have in the cartesian space. As mentioned above in Section 2.1, the hover state can be used as a point of work to approximate the helicopter's dynamics by linear equations of motion. Using this approximation, linearization around this state gives a wide enough range of linearity that is useful for control purposes. For this reason, this control is formed of decoupled PID controllers for each of the control variables described above.

The second controller is a velocity-based control responsible of generating the references for the attitude control. It is implemented using a PI configuration. The controller reads the state of the vehicle from the state estimator and gives references to the next level, but only to make lateral and longitudinal displacements. The third controller (position based control) is at the higher level of the system, and is designed to receive GPS coordinates. The control scheme allows different modes of operation, one of which is to take the helicopter to a desired position (position control). Once the UAV is hovering, the velocity based control is capable of receiving references to keep the UAV aligned with a selected target, and it leaves the stability of the aircraft to the most internal loop in charge of the attitude. Figure 6 shows the structure of the

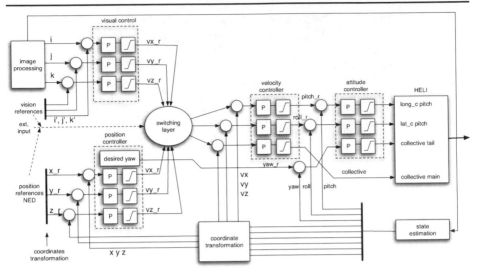

Fig. 6 Schematic flight control system. The inner velocity control loop is made of three cascade decoupled PID controllers. The outer position control loop can be externally switched between the visual based controller and the GPS based position controller. The former can be based on direct feature visual control or alternatively on visual estimated world positioning

flight control system with more details, and the communication interface described in Section 2.2, that is the key to integrate the visual reference as an external loop. Next subsection describes how this has been achieved.

4.2 Visual References Integration

The first step to design the control task in the image coordinates is to define the camera's model and the dynamics of a feature in the image, in order to construct a control law that properly represents the behavior of the task. Figure 7 shows the basic PinHole model of the camera, where $P^c(x, y, z)$ is a point in the camera coordinates system, and $p^c(i, j)^T$ denotes the projection of that point in the image plane π. Velocity of the camera can be represented with the vector $V = (v_x^c, v_y^c, v_z^c)^T$, while vector $\omega = (w_x^c, w_y^c, w_z^c)^T$ depicts the angular velocity. Considering that objects in the scene don't move, the relative velocity of a point in the world related to the camera's optical center can be expressed in this form:

$$\dot{P^c} = - \left(V + \omega \times P^c \right) \tag{9}$$

Using the well known Eq. 10 based on the camera calibration matrix that expresses the relationship between a point in the camera's coordinate system and its projection in the image plane, deriving Eq. 10 with respect to time, and replacing Eq. 9, it is possible to obtain a new Eq. 11 that describes a differential relation between the

Fig. 7 PinHole camera model to describe the dynamic model, where $P(x, y, z)$ is a point in the camera coordinates system, $p(i, j)^T$ represents the projection of that point in the image plane π and the vector $\omega = (w_x, w_y, w_z)^T$ is the angular velocity

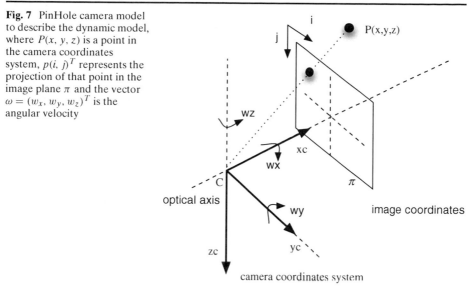

velocity of the projection of a point in the image and the velocity vector of the camera V and ω.

$$p^c = \mathbf{K} P^c \tag{10}$$

$$\dot{p^c} = -\mathbf{K}\left(V + \omega \times P^c\right) \tag{11}$$

Since the visual servoing task is designed to only make lateral and longitudinal displacements, and the camera is fixed looking to the front, it is possible to assume that the angular velocity is despicable because of the short range of motion of the pitch angles and the velocity constraint imposed to the system. Hence, Eq. 11 is reduced to this expression:

$$\dot{p^c} = \begin{bmatrix} \dfrac{di}{dt} \\ \dfrac{dj}{dt} \end{bmatrix} = - \begin{bmatrix} \dfrac{f}{x^c} & 0 \\ 0 & \dfrac{f}{x^c} \end{bmatrix} \begin{bmatrix} v_x^c \\ v_z^c \end{bmatrix} \tag{12}$$

This expression permits the introduction of the references described in Section 3 as a single measure, using the center of mass of the features or the patch tracked by the image processing algorithm, and using the velocity control module of the Flight Control System described above in this section.

5 Stereo Vision

This section shows a system to estimate the altitude and motion of an aerial vehicle using a stereo visual system. The system first detects and tracks interest points in the scene. The depth of the plane that contains the features is calculated matching features between left and right images and then using the disparity principle. Motion is recovered tracking pixels from one frame to the next one, finding its visual displacement and resolving camera rotation and translation by a least-square method [28].

5.1 Height Estimation

Height Estimation is performed on a stereo system using a first step to detect features in the environment with any of the technique mentioned in Section 3. This procedure is performed in each and every one of the stereo images.

As a second step, a correlation procedure is applied in order to find the correspondences between the two sets of features from the right and left images. Double check is performed by checking right against left, and then comparing left with right. The correlation stage is based on the ZNNC—zero mean normalized cross correlation—which offers good robustness against light and environmental changes [29].

Once the correspondence has been solved, considering an error tolerance, given that the correspondence is not perfect, and thanks to the fact that all pixels belong to the same plane, the stereo disparity principle is used to find the distance to the plane that contains the features. Disparity is inversely proportional to scene depth multiplied by the focal length (f) and baseline (b). The depth is computed using the expression for Z shown in Fig. 8.

Figure 9 shows the algorithm used to estimate the distance from the stereo system to the plane. In the helicopter, the stereo system is used in two positions. In the first one, the stereo system is looking down, perpendicular to ground, so that the estimated distance corresponds to the UAV altitude. In the second configuration, the stereo system is looking forward, and by so doing the estimated distance corresponds to the distance between the UAV and an object or feature.

Fig. 8 Stereo Disparity for aligned cameras with all pixel in the same plane. Stereo disparity principle is used to find the distance to the plane that contains the features

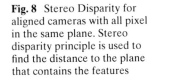

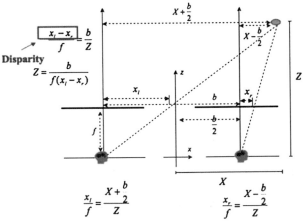

Fig. 9 Height estimation using the Harris Corner detector and ZNNC. Height is obtained employing the stereo disparity principle

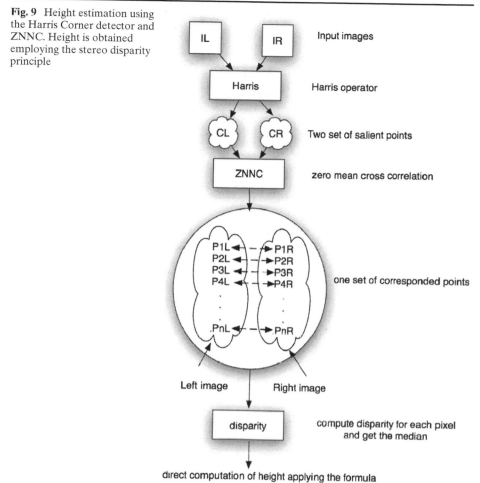

5.2 Motion Estimation

Motion estimation is performed using at a first stage the same technique used for feature correspondence between left and right corners: the zero mean normalized cross-correlation (ZNNC). Correlation is performed within a certain pixel distance from each other keeping those points in a correlation coefficient higher than 0.85. The motion problem estimation is done aligning two sets of points whose correspondence is known, and finding the rotation matrix and translation vector, i.e. 3D transformation matrix T that minimizes the mean-squares' objective function $\min_{R,t} \sum_N \| T P_{k-1} - P_k \|^2$. Problem can be solved using Iterative Closes Point (ICP) registration and motion parameter estimation using SVD. Assuming there are two sets of points which are called data and model: $P = \{p_i\}_1^{N_p}$ and $M = \{m_i\}_1^{N_m}$ respectively with $N_p \neq N_m$, whose correspondence is known. The problem is how to compute the rotation (R) and translation (t) producing the best possible alignment of P and M by relation them with the equation $M = RP + t$. Lets define the closest point in the model to a data point p as $cp(p) = \arg \min_{m \in M} \| m - p \|$. Then, the ICP step goes like this:

1. Compute the subset of closest points (CP), $y = \{m \in M \mid p \in P : m = cp(p)\}$
2. Compute the least-squares estimate of motion bringing P onto y:
   $(R, t) = \arg \min_{R,t} \sum_{i=1}^{N_p} \| y_i - Rp_i - t \|^2$
3. Apply motion to the data points, $P \leftarrow RP + t$
4. If the stopping criterion is satisfied, exit; else goto 1.

Calculating the rotation and the translation matrix using SVD can be summarized as follows: first, the rotation matrix is calculated using the centroid of the set of points. Centroid is calculated as $y_{c_i} = y_i - \bar{y}$ and $p_{c_i} = p_i - \bar{p}$, where $\bar{y} = \frac{1}{N_p} \sum_{N_p} cp(p_i)$ and $\bar{p} = \frac{1}{N_p} \sum_{N_p} p_i$. Then, rotation is found minimizing $\min_R \sum_{N_p} \| y_{c_i} - Rp_{c_i} \|^2$. This equation is minimized when trace (RK) is maximized with $K = \sum_{N_p} y_{c_i} p_{c_i}^T$. Matrix K is calculated using SVD as $K = VDU^T$. Thus, the optimal rotation matrix that maximizes the trace is $R = VU^T$. The optimal translation that aligns the centroid is $t = \bar{y} - P\bar{p}$.

Section 7.3, shows tests and applications' development using the stereo system and algorithms explained in this section.

6 Airborne Visual SLAM

This section presents the implementation of an aerial visual SLAM algorithm with monocular information. No prior information of the scene is needed for the proposed formulation. In this approach, no extra absolute or relative information, GPS or odometry are used. The SLAM algorithm is based on the features or corners' matching process using SURF features [30] or on the Harris Corner detector [18]. First, the formulation of the problem will be described. Then, the details of the Kalman filter will be explained and, finally, this section will end with the description of this approach's particularities.

6.1 Formulation of the Problem

The problem is formulated using state variables to describe and model the system. The state of the system is described by the vector:

$$X = [\mathbf{x}, \mathbf{s}_1, \mathbf{s}_2, \mathbf{s}_3, ...] \tag{13}$$

where $\mathbf{x}$ denotes the state of the camera and $\mathbf{s}_i$ represents the state of each feature. Camera state has 12 variables. The First six variables represent the position of the vehicle in iteration k and in the previous iteration. The Next six variables, vector $[p, q, r]$, represent the rotation at iteration k and k-1. Rotation is expressed using

Rodrigues' notation, which expresses a rotation around a vector with the direction of $\omega = [p, q, r]$ of an angle $\theta = \sqrt{p^2 + q^2 + r^2}$. The rotation matrix is calculated from this representation using

$$e^{\tilde{\omega}\theta} = I + \tilde{\omega}\sin(\theta) + \tilde{\omega}^2(1 - \cos(\theta)) \tag{14}$$

where I is the 3×3 identity matrix and $\tilde{\omega}$ denotes the antisymmetric matrix with entries

$$\tilde{\omega} = \begin{bmatrix} 0 & -r & q \\ r & 0 & -p \\ -q & p & 0 \end{bmatrix} \tag{15}$$

Therefore the state of the camera, not including the features, is composed by the following 12 variables,

$$\mathbf{x} = [x_k, x_{k-1}, y_k, y_{k-1}, z_k, z_{k-1}, p_k, p_{k-1}, q_k, q_{k-1}, r_k, r_{k-1}] \tag{16}$$

Another implementation of monocular SLAM uses quaternion to express the rotation [31]. The use of Rodrigues' notation, instead of quaternion, allows the reduction of the problem's dimension by only using three variables to represent the rotation.

Each feature is represented as a vector $[s_i]$ of dimension 6 using the inverse depth parametrization proposed by Javier Civera in [31]. This parametrization uses six parameters to define the position of a feature in a 3-Dimensional space. Each feature is defined by the position of a point (x_0, y_0, z_0) where the camera first saw the feature, the direction of a line based on that point and the inverse distance from the point to the feature along the line. This reference system allows the initialization of the features without any prior knowledge about the scene. This is important in exterior scenes where features with very different depths can coexist.

$$s_i = [x_0, y_0, z_0, \theta, \phi, \rho] \tag{17}$$

6.2 Prediction and Correction Stages

Extended Kalman filter (EKF) is used to implement the main algorithm loop, which has two stages: prediction and correction. In the prediction stage, uncertainty is propagated using the movement model. The correction stage uses real and predicted measurements to compute a correction to the prediction stage. Both stages need a precise description of the stochastic variables involved in the system.

There are mainly two approaches to implement this filter: extended Kalman filter and particle filter (FastSLAM). Both filters use the same formulation of the problem but have different approaches to the solution. The advantages of the Kalman filter are the direct estimation of the covariance matrix and the fact that it is a closed mathematical solution.

Its disadvantages are the increase of computational requirements as the number of features increase, the need of the model's linearization and the assumption of gaussian noise. On the other hand, particle filters can deal with non-linear, non-gaussian models, but the solution they provide depends on an initial random set of particles which can differ in each execution. Prediction stage is formulated using linear equations

$$\hat{X}_{k+1} = A \cdot X_k + B \cdot U_k$$
$$\hat{P}_{k+1} = A \cdot P_k \cdot A^T + Q \tag{18}$$

Where A is the transition matrix, B is the control matrix and Q is the model's covariance. Camera movement is modeled using a constant velocity model. Accelerations are included in a random noise component. For a variable n, which represents any of the position components (x, y, z) or the rotation components (p, q, r), we have:

$$n_{k+1} = n_k + v_k \cdot \Delta t \tag{19}$$

Where v_k is the derivative of n. We can estimate v_k as the differences in position,

$$n_{k+1} = n_k + \left(\frac{n_k - n_{k-1}}{\Delta t} \right) \Delta t = 2n_k - x_{n-1} \tag{20}$$

Feature movement is considered constant and therefore is modeled by an identity matrix. Now, full state model can be constructed

$$
\begin{bmatrix}
x_{k+1} \\
x_k \\
y_{k+1} \\
y_k \\
z_{k+1} \\
z_k \\
r_{k+1} \\
r_k \\
p_{k+1} \\
p_k \\
q_{k+1} \\
q_k \\
s_{1,k+1} \\
\cdots
\end{bmatrix}
=
\begin{bmatrix}
2 & -1 & & & & & & \\
1 & 0 & & & & & & \\
& & 2 & -1 & & & & \\
& & 1 & 0 & & & & \\
& & & & 2 & -1 & & \\
& & & & 1 & 0 & & \\
& & & & & & 2 & -1 \\
& & & & & & 1 & 0 \\
& & & & & & & & 2 & -1 \\
& & & & & & & & 1 & 0 \\
& & & & & & & & & & 2 & -1 \\
& & & & & & & & & & 1 & 0 \\
& & & & & & & & & & & & \mathbf{I} \\
& & & & & & & & & & & & & \ddots
\end{bmatrix}
\begin{bmatrix}
x_k \\
x_{k-1} \\
y_k \\
y_{k-1} \\
z_k \\
z_{k-1} \\
r_k \\
r_{k-1} \\
p_k \\
p_{k-1} \\
q_k \\
q_{k-1} \\
s_{1,k} \\
\cdots
\end{bmatrix}
\tag{21}
$$

Correction stage uses a non-linear measurement model. This model is the pin-hole camera model. The formulation of the Extended Kalman Filter in this scenario is

$$K_k = \hat{P}_k \cdot J^T \left(J \cdot P \cdot J^T + R \right)^{-1}$$
$$X_k = \hat{X}_k + K_k \cdot \left(Z_k - H\left(\hat{X}_k\right) \right)$$
$$P_k = \hat{P}_k - K_k \cdot J \cdot \hat{P}_k \tag{22}$$

where Z_k is the measurement vector, $H(X)$ is the non-linear camera model, J is the jacobian of the camera model and K_k is the Kalman gain.

The movement of the system is modeled as a solid with constant motion. Acceleration is considered a perturbation to the movement. A pin-hole camera model is used as a measurement model.

$$
\begin{bmatrix} \lambda u \\ \lambda v \\ \lambda \end{bmatrix} = \begin{bmatrix} f & 0 & 0 \\ 0 & f & 0 \\ 0 & 0 & 1 \end{bmatrix} \cdot [R|T] \cdot \begin{bmatrix} x_w \\ y_w \\ z_w \\ 1 \end{bmatrix}
\tag{23}
$$

where u and v are the projected feature's central coordinates and λ is a scale factor. Distortion is considered using a four parameter model (k1, k2, k3, k4)

$$
r^2 = u^2 + v^2
$$

$$
C_{dist} = 1 + k_0 r^2 + k_1 r^4
$$

$$
x_d = u \cdot C_{dist} + k_2 (2u \cdot v) + k_3 \left(r^2 + 2u^2\right)
$$

$$
y_d = v \cdot C_{dist} + k_2 \left(r^2 + 2v^2\right) + k_3 (2u \cdot v)
\tag{24}
$$

State error covariance matrix is initialized in a two part process. First, elements related to the position and orientation of the camera, **x**, are initialized as zero or as a diagonal matrix with very small values. This represents that the position is known, at the first instant, with very low uncertainty. The initialization of the values related to the features, s_i, must be done for each feature seen for first time. This initialization is done using the results from [31]:

$$
\mathbf{P}_{k|k}^{new} = J \begin{bmatrix} \mathbf{P}_{k|k} & & \\ & \mathbf{R}_i & \\ & & \sigma_\rho^2 \end{bmatrix} J^T
\tag{25}
$$

where

$$
\mathbf{J} = \begin{bmatrix} I & & 0 & 0 & \\ \frac{\partial s}{\partial \mathbf{xyz}} & \frac{\partial s}{\partial \mathbf{pqr}} & 0 & 0 & \cdots & \frac{\partial s}{\partial x_d, y_d} & \frac{\partial s}{\partial \rho_0} \end{bmatrix}
\tag{26}
$$

$$
\frac{\partial s}{\partial \mathbf{xyz}} = \begin{bmatrix} 1 & 0 & 0 \\ 0 & 1 & 0 \\ 0 & 0 & 1 \\ 0 & 0 & 0 \\ 0 & 0 & 0 \\ 0 & 0 & 0 \end{bmatrix} ; \quad \frac{\partial s}{\partial \mathbf{pqr}} = \begin{bmatrix} 0 & 0 & 0 \\ 0 & 0 & 0 \\ 0 & 0 & 0 \\ \frac{\partial \theta}{\partial p} & \frac{\partial \theta}{\partial q} & \frac{\partial \theta}{\partial r} \\ \frac{\partial \phi}{\partial p} & \frac{\partial \phi}{\partial q} & \frac{\partial \phi}{\partial r} \\ \frac{\partial \phi}{\partial p} & \frac{\partial \phi}{\partial q} & \frac{\partial \phi}{\partial r} \\ 0 & 0 & 0 \end{bmatrix} ; \quad \frac{\partial s}{\partial x_d, y_d} = \begin{bmatrix} 0 & 0 \\ 0 & 0 \\ 0 & 0 \\ \frac{\partial \theta}{\partial x_d} & \frac{\partial \theta}{\partial y_d} \\ \frac{\partial \phi}{\partial x_d} & \frac{\partial \phi}{\partial y_d} \\ 0 & 0 \end{bmatrix} ; \quad \frac{\partial s}{\partial \rho_0} = \begin{bmatrix} 0 \\ 0 \\ 0 \\ 0 \\ 0 \\ 1 \end{bmatrix}
\tag{27}
$$

Taking into account that a robust feature tracking and detection is a key element in the system, a Mahalanobis' test is used in order to improve the robustness of feature matching. The filter is implemented using Mahalanobis' distance between the predicted feature measurement and the real measurement. Mahalanobis' distance weighs Euclidean distance with the covariance matrix. This distance is the input to a χ^2 test which rejects false matches.

$$
(Z - J \cdot X)^t \cdot C^{-1}(Z - J \cdot X) > \chi_n^2
\tag{28}
$$

where

$$C = H \cdot P \cdot H^T + R \tag{29}$$

Finally, it should be noted that the reconstruction scale is an unobservable system state. This problem is dealt with using inverse depth parametrization [32], which avoids the use of initialization features of known 3D positions. This permits the use of the algorithm in any video sequence. Without these initialization features, the problem becomes dimensionless. The scale of the system can be recovered using the distance between two points or the position between the camera and one point. Computational cost is dependant on the number of features in the scene, and so an increase in the scene's complexity affects processing time in a negative way. Robust feature selection and matching are very important to the stability of the filter and a correct mapping. Experiments carried out successfully were made offline on sequences taken from the UAV.

7 Experimental Application and Tests

7.1 Visual Tracking Experiments

Tracking algorithms are fundamental to close the vision control loop in order to give an UAV the capability to follow objects. Hence, it is important to ensure the reliability of the tracker. Some experiments were conducted on images taken on test flights. Such experiments, where interest points were extracted with the Harris algorithm and tracked with the Lukas–Kanade algorithm, have proven to be fast enough so as to close the control loop at 17 Hz. However, if there are too many features selected to represent an object, the algorithm's speed slows down because of the calculation of the image derivatives.

SIFT features are very robust and rely on the advantage that the matching process does not depend on the proximity of two consecutive frames. On the other hand, the computational cost of the extraction is expensive. For that reason, they are suitable for visual servoing only if the displacements of the helicopter are forced to be very slow in order to avoid instabilities when closing the loop.

Tracking based on appearance proves to be very fast and reliable for acquired sequences at frame rates above 25 fps. This procedure is very sensitive to abrupt changes in the position of the tracked patch as long as the number of parameters of the motion model is higher than 3. This can be solved using stacks of trackers, each of which must have a different warping function that provides an estimation of the parameter to the next level of the stack. Simple warping functions give an estimation of more complex parameters. In the case of a simple tracker the translation-only warping function is the most stable one. Figure 10a shows the evolution of the parameters in a sequence of 1,000 images, and Fig. 10b the SSD error between the template image and the warped patch for each image.

7.2 Visual Servoing Experiments

The basic idea of visual servoing is to control the position of the helicopter based on an error in the image, or in a characteristic extracted from the image. If the

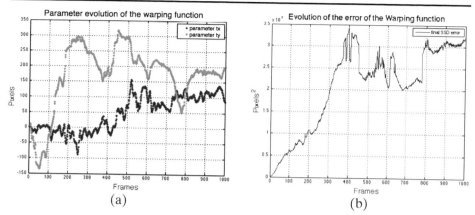

Fig. 10 Evolution of the translation parameter during the tracking process of a patch along 1,000 frames (**a**). **b** shows the SSD error of warped patch with respect to the template

control error is in the image plane, the measure of the error is a vector (in pixels) that represents the distance from the image's center to the feature's position. Figure 11 shows the basic idea of the error and the 2D visual servoing. In this sense, there are two ways to use this error in different contexts. One approach is to track features that are static in the scene. In this case, the control tries to move the UAV to align the feature's position with the image's center by moving the helicopter in the space.

Vision-based references are translated into helicopter displacements based on the tracked features. Velocity references are used to control the UAV, so that when the feature to track changes— as happens, for example, when another window of a building is chosen—velocity references change in order to align the UAV with the window.

The displacement of the helicopter when it tries to align with the feature being tracked is displayed in Fig. 12a. Vertical and lateral displacements of the helicopter are the consequence of the visual references generated from the vertical and horizontal positions of the window in the image. Figure 12b shows the displacement of the helicopter when the window above displayed was tracked, and Fig. 13 shows the velocity references when another window is chosen.

Fig. 11 Error measure in 2D visual servoing consists in the estimation of the distance of the reference point to the image's center

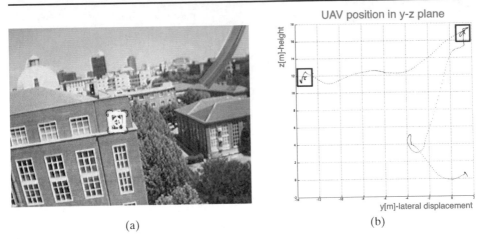

(a)   (b)

Fig. 12 Window being tracked during a visual servoing task (**a**), in which the UAV's vertical and lateral displacements are controlled by the visual control loop in order to fix the window in the center of the image, while the approaching movement is controlled by the GPS position controller. **b** Shows UAV vertical and lateral positions during the visual controlled flight. After taking off, the UAV moves to two positions (*marked with the red rectangles*) in order to consecutively track two external visual references that consist of two different windows

Another possible scenario is to keep the UAV hovering and to track moving objects in the scene. Experiments have been conducted successfully in order to proof variation of the method with good results. Control of the camera's Pan-Tilt Platform using 2D image servoing tries to keep a moving object in the image's center. In this case, position references are used instead of velocity in order to control the camera's pan and tilt positions. Figure 14 shows a car carrying a poster being tracked by moving the camera's platform.

Fig. 13 Velocity references change when a new feature is selected, in this case when another window is selected as shown in Fig. 12. Visual control takes the feature to the image center

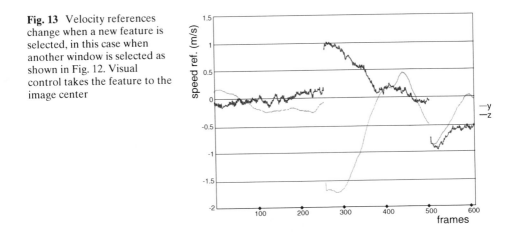

Fig. 14 Tracking of moving object. Servoing is perform on the pan-tilt platform. Notice that velocity in the cartesian coordinates is 0.0 (each component is printed on the image) since the UAV is hovering. Tracking is performed using corner features as explained in Section 3.2

7.3 Height and Motion Estimation Using a Stereo System

Stereo tests are made using a Firewire stereo system camera onboard the UAV. In these experiments, the helicopter is commanded to fly autonomously following a given trajectory while the onboard stereo vision algorithm is running. The experiments find the correlation between the stereo visual estimation and the onboard helicopter state given by its sensor suite. Figure 15 shows the results of one flight trial in which the longitudinal displacement (X), lateral displacement (Y), altitude (H) and relative orientation are estimated. Altitude is computed negative since the helicopter's body frame is used as a reference system. Each estimation is correlated with its similar value taken from the onboard helicopter state, which uses an EKF to fuse onboard sensors. Table 2 shows the error analysis based on the mean square error of the visual estimation and the helicopter's state. Four measures of the mean squared error are used: the error vision-GPS Northting (MSE_N^V), the error vision-GPS Easting (MSE_E^V), the error vision-yaw (MSE_ψ^V) and the error vision-altitude (MSE_H^V).

7.4 Power Lines Inspection

Besides visual servoing and image tracking applications, other experiments have been conducted to achieve object recognition in inspection tasks. Major contributions and successful tests were obtained in power lines' inspection. The objective of the application developed at the computer vision group is to identify powered lines and electrical isolators. The methodology that has been employed is based on the Hough transform and on Corner detectors that find lines in the image that are associated with the catenary curve formed by the hanging wire. Interest points are used to locate the isolator. Once both components are detected in the image, tracking can be initiated to make close up shots with the appropriate resolution needed for expert inspection and detection of failures. Figure 16 shows images of the UAV

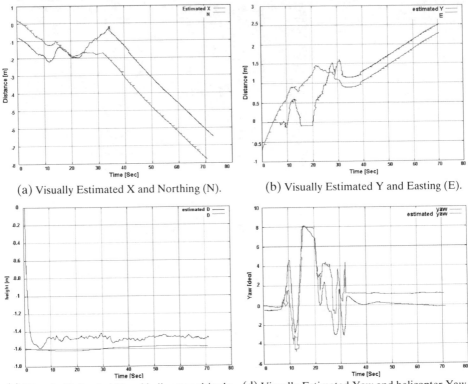

(a) Visually Estimated X and Northing (N).

(b) Visually Estimated Y and Easting (E).

(c) Visually Estimated H and helicopter altitude.

(d) Visually Estimated Yaw and helicopter Yaw.

Fig. 15 Results using a stereo system. Four parameters are estimated for this experiment: the longitudinal displacement (X) (**a**), the lateral displacement (Y) (**b**), altitude (H) (**c**) and relative orientation (yaw) (**d**)

approaching a power line while in the sub-image the onboard camera displays the detection of the line and the isolator.

Stereo System has also been used to estimate the UAV distance and altitude with respect to power lines. In these tests, the line is detected using the Hough Transform. If the camera's angles, stereo system calibration and disparity are known, it is possible to determine the position of the helicopter relative to the power line. Some tests using the Stereo system onboard the helicopter were carried out to obtain the distance to the power line from the helicopter. The power Line is detected using Hough transform in both images. In this test, the helicopter was initially 2 m below the power line. Afterwards, it rises to be at the same altitude of the cable and then

Table 2 Error analysis for the helicopter's experimental trials	Exp.	Test
	MSE_N^V **m**	1.0910
	MSE_E^V **m**	0.4712
	MSE_ψ^V **deg**	1.7363
	MSE_H^V **m**	0.1729

Fig. 16 Power line and Isolator detection using the UAV vision system

it returns to its original position. Figure 17 shows the distance and height estimated from the UAV to the power line during this test. Additional tests can be seen on the Colibri Project's Web Page [33].

7.5 Mapping and Positioning using Visual SLAM

The SLAM algorithm explained in Section 6 is used in a series of image sequences of trajectories around a 3D scene that were performed flying in autonomous mode navigation based on way points and desired heading values. The scene is composed of many objects, including a grandstand, a van and many other elements, and also of a series of marks feasible for features and corners' detection. For each flight test, a 30 fps image sequence of the scene was obtained, associating the UAV attitude information for each one. That includes the GPS position, IMU data (Heading, body frame angles and displacement velocities) and the helicopter's position, estimated by the Kalman filter on the local plane with reference to takeoff point. Figure 18 shows a reconstruction of one flight around one scene test.

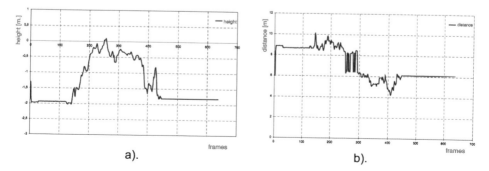

a).   b).

Fig. 17 Distance and height estimation to the power lines using a stereo system onboard the UAV

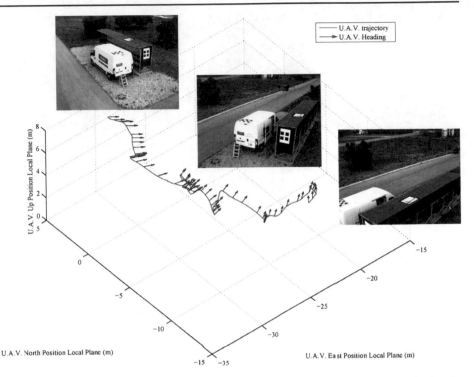

Fig. 18 Three-dimensional flight trajectory and camera position reconstruction obtained using the flightlog data. The *blue line* depicts the translational movement and the *red arrows* represent the heading direction of the camera (pitch and yaw angles). *Superimposed images* show the different perspectives obtained during the flight sequence around the semi-structured scene

Results for tests using a tracking algorithm for scene elements are shown on Fig. 19a. Reconstructed features are shown as crosses. In the figure, some reference planes were added by hand in order to make interpretation easier. Figure 19b shows an image from the sequence used in this test.

Results show that the reconstruction has a coherent structure but that the scale of the reconstruction is function of the initialization values. The scale can be recovered using the distance between two points or the positions of one point and the camera.

The camera movement relative to the first image is compared with the real flight trajectory. For this, the (x, y, z) axis on the camera plane are rotated so that they are coincident with the world reference plane used by the UAV. The heading or yaw angles (ψ) and the Pitch angle (θ) of the helicopter, in the first image of the SLAM sequence, define the rotational matrix used to align the camera and UAV frames.

The displacement values obtained using SLAM are rotated and then scaled to be compared with the real UAV trajectory. Figure 20 shows the UAV and SLAM trajectories and the medium square error (MSE) between real flight and SLAM displacement for each axe. The trajectory adjusts better to the real flight as the features reduce their uncertainty, because the more images are processed, more measurements refine features estimation.

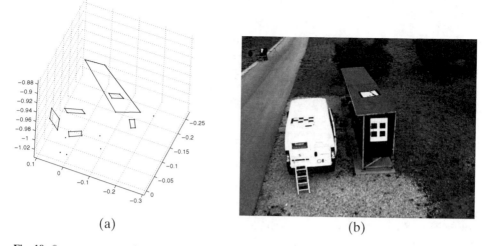

(a)   (b)

Fig. 19 Scene reconstruction. The *upper figure* shows reconstructed points from the scene shown in the *lower figure*. *Points* are linked manually with lines to ease the interpretation of the figure

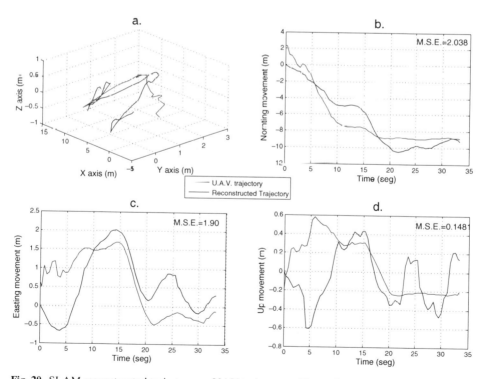

Fig. 20 SLAM reconstructed trajectory vs. UAV trajectory. **a** Three-dimensional flight, **b** north axe in meters, **d** east axe in meters, **c** altitude in meters. The reconstructed trajectory adjusts best to the real flight as soon as more images are processed and the uncertainty of the features is thus reduced

8 Conclusions

This paper dealt with the researches, results and discussion of the use of several techniques of computer vision onboard an UAV. These computer vision techniques are not merely used for acquiring environmental visual information that can be used afterwards by off-line processing. That's why the paper also shows how computer vision can play an important role on-line during the flight itself in order to acquire the adequate sequences necessary to actively track targets (fixed or moving ones) and to guide and control flight trajectories.

Image processing algorithms are very important, and are often designed to detect and track objects along the sequences, whether key points are extracted by the algorithm itself or are externally determined visual targets. Successful, wide spread algorithms onboard an UAV have test bed challenges and thus provide a source of inspiration for their constant improvement and for achieving their better robustness. Some of those test bed challenges are the non-structured and changing light conditions, the highly vibrating and quick and sharp movements, and on-line requirements when necessary.

Some improvements have been presented and tested in the following two types of image processing algorithms: feature tracking and appearance-based tracking, due to the above mentioned characteristics. When using the SIFT key point detector, the algorithm reduces and classifies the key points for achieving a more robust and quick tracking as stated in Section 3. When tracking a whole visual target, an ICA based algorithm is used in a multi-scale hierarchical architecture that makes it robust for scaling. In both type of algorithms, a Kalman filter has been implemented in order to improve the consistence of the features and targets' movements within the image plane, a feat that is particularly relevant in quick changing sequences, as stated in Section 3.3.

The filtered outputs of the image processing algorithms are the visual measurements of the external references that, when compared to their desired position, are introduced in a decoupled position control structure that generates the velocity references in order to control the position of the UAV according to those external visual references. Depending on the type of information extracted by the image processing algorithms (i.e. bi-dimensional translation, rotation, 3D measurements, among others), the UAV's position and orientation control can be a mix of visual based control for some UAV coordinates and GPS based control for some others. A Kalman filter can also be computed in future developments to produce unified UAV estimation and control based on visual, GPS, and inertial information.

This paper also shows that it is possible to obtain robust and coherent results using Visual SLAM for 3D mapping and positioning in vague structured outdoor scenes from a mini UAV. The SLAM algorithm has been implemented using only visual information without considering any odometric or GPS information. Nonetheless, this information has been later used in order to compare and evaluate the obtained results. The state of the system comprises a 12 variable array (position, orientation and their rates), where the inverse depth parametrization has been used in order to avoid the initialization of the distances to the detected visual features, that otherwise becomes a drawback when using SLAM outdoors in unknown environments. The rest of the state array is made up of the tracked features, being ten the minimum allowed number. The prediction stage in EKF has been modeled considering constant velocity for both the position-orientation coordinates and the feature movements in the image plane. The correlation stage in the EKF uses a non-linear camera model that includes a pin-hole distortion model for the sake of more accurate results. Within the implemented SLAM algorithm the Mahalanobis' distance is used to discharge far away matched pairs that can otherwise distort the results.

Based on the results of our work, we conclude that the UAV field has reached an important stage of maturity in which the possibility of using UAVs in civilian applications is now imaginable and in some cases attainable. We have experimentally demonstrated several capabilities that an autonomous helicopter can have by using visual information such as navigation, trajectory planning and visual servoing. The successful implementation of all these algorithms confirms the necessity of dotting UAVs with additional functionalities when tasks like outdoor structures' inspection and object tracking are required.

Our current work is aimed at increasing these capabilities using different visual information sources like catadioptric systems and multiple view systems, and extending them to 3D image-based visual servoing, where the position and orientation of the object will be used to visually conduct the helicopter. The challenge is to achieve real-time image processing and tracking algorithms to reduce the uncertainty of the measure. The field of computer vision for UAVs can be considered as a promising area for investing further research for the benefit of the autonomy and applicability of this type of aerial platforms, considering that reliability and safety have become major research issues of our community.

Acknowledgements   The work reported in this paper is the product of several research stages at the Computer Vision Group Universidad Politécnica de Madrid. The authors would like to thank Jorge León for supporting the flight trials and the I.A. Institute—CSIC for collaborating in the flights' consecution. We would also like to thank Enrique Munoz Corral and Luis Baumela for helping us to understand and put into practice algorithms to track planar objects. This work has been sponsored by the Spanish Science and Technology Ministry under grants CICYT DPI2004-06624, CICYT DPI2000-1561-C02-02 and MICYT DPI2007-66156, and by the Comunidad Autónoma de Madrid under grant SLAM visual 3D.

References

1. Puri, A., Valavanis, K.P., Kontitsis, M.: Statistical profile generation for traffic monitoring using real-time UAV based video data. In: Control and Automation, 2007. MED '07. Mediterranean Conference on, MED, pp. 1–6 (2007)
2. Nikolos, I.K., Tsourveloudis, N.C., Valavanis, K.P.: Evolutionary algorithm based path planning for multiple UAV cooperation. In: Advances in Unmanned Aerial Vehicles, Intelligent Systems, Control and Automation: Science and Engineering, pp. 309–340. Springer, The Netherlands (2007)
3. Nikolos, I.K., Tsourveloudis, N.C., Valavanis, K.P.: A UAV vision system for airborne surveillance. In: Robotics and Automation, 2004. Proceedings. ICRA '04. 2004 IEEE International Conference on, pp. 77–83. New Orleans, LA, USA (2004), May
4. Nikolos, I.K., Tsourveloudis, N.C., Valavanis, K.P.: Multi-UAV experiments: application to forest fires. In: Multiple Heterogeneous Unmanned Aerial Vehicles, Springer Tracts in Advanced Robotics, pp. 207–228. Springer, Berlin (2007)
5. Green, W., Oh, P.Y.: The integration of a multimodal mav and biomimetic sensing for autonomous flights in near-earth environments. In: Advances in Unmanned Aerial Vehicles, Intelligent Systems, Control and Automation: Science and Engineering, pp. 407–430. Springer, The Netherlands (2007)

6. Belloni, G., Feroli, M., Ficola, A., Pagnottelli, S., Valigi, P.: Obstacle and terrain avoidance for miniature aerial vehicles. In: Advances in Unmanned Aerial Vehicles, Intelligent Systems, Control and Automation: Science and Engineering, pp. 213–244. Springer, The Netherlands (2007)
7. Dalamagkidis, K., Valavanis, K.P., Piegl, L.A.: Current status and future perspectives for unmanned aircraft system operations in the US. In: Journal of Intelligent and Robotic Systems, pp. 313–329. Springer, The Netherlands (2007)
8. Long, L.N., Corfeld, K.J., Strawn, R.C.: Computational analysis of a prototype martian rotorcraft experiment. In: 20th AIAA Applied Aerodynamics Conference, number AIAA Paper 2002–2815, Saint Louis, MO, USA. Ames Research Center, June–October 22 (2001)
9. Yavrucuk, I., Kanan, S., Kahn, A.D.: Gtmars—flight controls and computer architecture. Technical report, School of Aerospace Engineering, Georgia Institute of Technology, Atlanta (2000)
10. Buenaposada, J.M., Munoz, E., Baumela, L.: Tracking a planar patch by additive image registration. In: Proc. of International Workshop, VLBV 2003, vol. 2849 of LNCS, pp. 50–57 (2003)
11. Miller, R., Mettler, B., Amidi, O.: Carnegie mellon university's 1997 international aerial robotics competition entry. In: International Aerial Robotics Competition (1997)
12. Montgomery, J.F.: The usc autonomous flying vehicle (afv) project: Year 2000 status. Technical Report IRIS-00-390, Institute for Robotics and Intelligent Systems Technical Report, Los Angeles, CA, 90089-0273 (2000)
13. Saripalli, S., Montgomery, J.F., Sukhatme, G.S.: Visually-guided landing of an unmanned aerial vehicle. IEEE Trans. Robot Autom. **19**(3), 371–381, June (2003)
14. Mejias, L.: Control visual de un vehiculo aereo autonomo usando detección y seguimiento de características en espacios exteriores. PhD thesis, Escuela Técnica Superior de Ingenieros Industriales. Universidad Politécnica de Madrid, Spain, December (2006)
15. Mejias, L., Saripalli, S., Campoy, P., Sukhatme, G.: Visual servoing approach for tracking features in urban areas using an autonomous helicopter. In: Proceedings of IEEE International Conference on Robotics and Automation, pp. 2503–2508, Orlando, FL, May (2006)
16. Mejias, L., Saripalli, S., Sukhatme, G., Campoy, P.: Detection and tracking of external features in a urban environment using an autonomous helicopter. In: Proceedings of IEEE International Conference on Robotics and Automation, pp. 3983–3988, May (2005)
17. Mejias, L., Saripalli, S., Campoy, P., Sukhatme, G.: Visual servoing of an autonomous helicopter in urban areas using feature tracking. J. Field Robot. **23**(3–4), 185–199, April (2006)
18. Harris, C.G., Stephens, M.: A combined corner and edge detection. In: Proceedings of the 4th Alvey Vision Conference, pp. 147–151 (1988)
19. Lowe, D.G.: Distinctive image features from scale-invariant keypoints. Int. J. Computer Vision **60**(2), 91–110 (2004)
20. Duda, R.O., Hart, P.E.: Use of the hough transformation to detect lines and curves in pictures. Commun. ACM **15**(1), 11–15 (1972)
21. Canny, J.: A computational approach to edge detection. IEEE Trans. Pattern Anal. Machine Intel. **8**(6), 679–698, November (1986)
22. Feldman, G., Sobel, I.: A 3×3 isotropic gradient operator for image processing. Presented at a talk at the Stanford Artificial Project (1968)
23. Mejías, L., Mondragón, I., Correa, J.F., Campoy, P.: Colibri: Vision-guided helicopter for surveillance and visual inspection. In: Video Proceedings of IEEE International Conference on Robotics and Automation, Rome, Italy, April (2007)
24. Lucas, B.D., Kanade, T.: An iterative image registration technique with an application to stereo vision. In: Proc. of the 7th IJCAI, pp. 674–679, Vancouver, Canada (1981)
25. Beis, J.S., Lowe, D.G.: Shape indexing using approximate nearest-neighbour search in high-dimensional spaces. In: CVPR '97: Proceedings of the 1997 Conference on Computer Vision and Pattern Recognition (CVPR '97), p. 1000. IEEE Computer Society, Washington, DC, USA (1997)
26. Fischer, M.A., Bolles, R.C.: Random sample concensus: a paradigm for model fitting with applications to image analysis and automated cartography. Commun. ACM **24**(6), 381–395 (1981)
27. Baker, S., Matthews, I.: Lucas-kanade 20 years on: A unifying framework: Part 1. Technical Report CMU-RI-TR-02-16, Robotics Institute, Carnegie Mellon University, Pittsburgh, PA, July (2002)
28. Mejias, L., Campoy, P., Mondragon, I., Doherty, P.: Stereo visual system for autonomous air vehicle navigation. In: 6th IFAC Symposium on Intelligent Autonomous Vehicles (IAV 07), Toulouse, France, September (2007)

29. Martin, J., Crowley, J.: Experimental comparison of correlation techniques. Technical report, IMAG-LIFIA, 46 Av. Félix Viallet 38031 Grenoble, France (1995)
30. Bay, H., Tuytelaars, T., Van Gool, L.: SURF: Speeded Up Robust Features. In: Proceedings of the Ninth European Conference on Computer Vision, May (2006)
31. Montiel, J.M.M., Civera, J., Davison, A.J.: Unified inverse depth parametrization for monocular slam. In: Robotics: Science and Systems (2006)
32. Civera, J., Davison, A.J., Montiel, J.M.M.: Dimensionless monocular slam. In: IbPRIA, pp. 412–419 (2007)
33. COLIBRI. Universidad Politécnica de Madrid. Computer Vision Group. COLIBRI Project. http://www.disam.upm.es/colibri (2005)

Vision-Based Odometry and SLAM for Medium and High Altitude Flying UAVs

F. Caballero · L. Merino · J. Ferruz · A. Ollero

Originally published in the Journal of Intelligent and Robotic Systems, Volume 54, Nos 1–3, 137–161.
© Springer Science + Business Media B.V. 2008

Abstract This paper proposes vision-based techniques for localizing an unmanned aerial vehicle (UAV) by means of an on-board camera. Only natural landmarks provided by a feature tracking algorithm will be considered, without the help of visual beacons or landmarks with known positions. First, it is described a monocular visual odometer which could be used as a backup system when the accuracy of GPS is reduced to critical levels. Homography-based techniques are used to compute the UAV relative translation and rotation by means of the images gathered by an onboard camera. The analysis of the problem takes into account the stochastic nature of the estimation and practical implementation issues. The visual odometer is then integrated into a simultaneous localization and mapping (SLAM) scheme in order to reduce the impact of cumulative errors in odometry-based position estimation

This work is partially supported by the AWARE project (IST-2006-33579) funded by the European Commission, and the AEROSENS project (DPI-2005-02293) funded by the Spanish Government.

F. Caballero (✉) · J. Ferruz · A. Ollero
University of Seville, Seville, Spain
e-mail: caba@cartuja.us.es

J. Ferruz
e-mail: ferruz@cartuja.us.es

A. Ollero
e-mail: aollero@cartuja.us.es

L. Merino
Pablo de Olavide University, Seville, Spain
e-mail: lmercab@upo.es

approaches. Novel prediction and landmark initialization for SLAM in UAVs are presented. The paper is supported by an extensive experimental work where the proposed algorithms have been tested and validated using real UAVs.

Keywords Visual odometry · Homography · Unmanned aerial vehicles · Simultaneous localization and mapping · Computer vision

1 Introduction

Outdoor robotics applications in natural environments sometimes require different accessibility capabilities than the capabilities provided by existing ground robotic vehicles. In fact, in spite of the progress in the development of unmanned ground vehicles along the last 20 years, navigating in unstructured natural environments still poses significant challenges. The existing ground vehicles have inherent limitations to reach the desired locations in many applications. The characteristics of the terrain and the presence of obstacles, together with the requirement of fast response, may represent a major drawback to the use of any ground locomotion system. Thus, in many cases, the use unmanned aerial vehicles (UAVs) is the only effective way to reach the target to get information or to deploy instrumentation.

In the last ten years UAVs have improved their autonomy both in energy and information processing. Significant achievements have been obtained in autonomous positioning and tracking. These improvements are based on modern satellite-based position technologies, inertial navigation systems, communication and control technologies, and image processing. Furthermore, new sensing and processing capabilities have been implemented on-board the UAVs. Thus, today we can consider some UAVs as intelligent robotic systems integrating perception, learning, real-time control, situation assessment, reasoning, decision-making and planning capabilities for evolving and operating in complex environments.

In most cases, UAVs use the global position system (GPS) to determine their position. As pointed out in the Volpe Report [42], the accuracy of this estimation directly depends on the number of satellites used to compute the position and the quality of the signals received by the device; radio effects like multi-path propagation could cause the degradation in the estimation. In addition, radio frequency interferences with coexisting devices or jamming could make the position estimation unfeasible.

These problems are well known in robotics. Thus, odometry is commonly used in terrestrial robots as a backup positioning system or in sensor data fusion approaches. This local estimation allows temporally managing GPS faults or degradations. However, the lack of odometry systems in most aerial vehicles can lead to catastrophic consequences under GPS errors; incoherent control actions could be commanded to the UAV, leading to crash and the loss of valuable hardware. Moreover, if full autonomy in GPS-less environments is considered, then the problem of simultaneous localization and mapping (SLAM) should be addressed.

If small UAVs are considered, their low payload represents a hard restriction on the variety of devices to be used for odometry. Sensors like 3D or 2D laser scanners are too heavy and have an important dependence to the UAV distance to the ground. Although there exist small devices for depth sensing, their range is usually shorter than 15 m. Stereo vision systems have been successfully applied to low/medium size UAVs due to its low weight and versatility [4, 9, 18], but the rigid distance between the two cameras limits the useful altitude range.

Monocular vision seems to offer a good solution in terms of weight, accuracy and scalability. This paper proposes a monocular visual odometer and vision-based localization methods to act as backup systems when the accuracy of GPS is reduced to critical levels. The objective is the development of computer vision techniques for the computation of the relative translation and rotation, and for the localization of the vehicle based on the images gathered by a camera on-board the UAV. The analysis of the problem takes into account the stochastic nature of the estimation and practical implementation issues.

The paper is structured as follows. First, related work in vision based localization for UAVs is detailed. Then, a visual odometer based on frame-to-frame homographies is described, together with a robust method for homography computation. Later, the homography-based odometry is included in a SLAM scheme in order to overcome the error accumulation present in odometric approaches. The proposed SLAM approach uses the information provided by the odometer as main prediction hypothesis and for landmark initialization. Finally, conclusions and lessons learned are described.

1.1 Related Work

One of the first researches on vision applied to UAV position estimation starts in the ninetics at the Carnegie-Mellon University (CMU). In [1], it is described a vision-based odometer that allowed to lock the UAV to ground objects and sense relative helicopter position and velocity in real time by means of stereo vision. The same visual tracking techniques, combined with inertial sensors, were applied to autonomous take off, following a prescribed trajectory and landing. The CMU autonomous helicopter also demonstrated autonomous tracking capabilities of moving objects by using only on-board specialized hardware.

The topic of vision-based autonomous landing of airborne systems has been actively researched [30]. In the early nineties, Dickmanns and Schell [13] presented some results of the possible use of vision for landing an airplane. Systems based on artificial beacons and structured light are presented [44, 45]. The BEAR project at Berkeley is a good example of vision systems for autonomous landing of UAVs. In this project, vision-based pose estimation relative to a planar landing target and vision-based landing of an aerial vehicle on a moving deck have been researched [36, 40]. A technique based on multiple view geometry is used to compute the real motion of one UAV with respect to a planar landing target. An artificial target allows to establish quick matches and to solve the scale problem.

Computer vision has also been proposed for safe landing. Thus, in [15], a strategy and an algorithm relying on image processing to search the ground for a safe landing spot is presented. Vision-based techniques for landing on a artificial helipad of known shape are also presented in [34, 35], where the case of landing on a slow moving helipad is considered. In [37], the landing strategies of bees are used to devise a vision system based on optical flow for UAVs.

Corke et. al [9] have analyzed the use of stereo vision for height estimation in small size helicopters. In Georgia Tech, vision-based aided navigation for UAVs has been considered. Thus, in [43] the authors present an Extended Kalman Filter approach that combines GPS measurements with image features obtained from a known artificial target for helicopter position estimation.

In a previous work [5], the authors present a visual odometer for aerial vehicles using monocular image sequences, but no error estimation is provided by the algorithm, and the approach is limited to planar scenes. In [6], it is shown how a mosaic can be used in aerial vehicles to partially correct the drift associated to odometric approaches. This technique is extended in [7] with a minimization process that allows to improve the spatial consistency of the online built mosaic. Recently, in [8] the authors propose a visual odometer to compensate GPS failures. Image matching with geo-referenced aerial imagery is proposed to compensate the drift associated to odometry.

Although vision-based SLAM has been widely used in ground robots and has demonstrated its feasibility for consistent perception of the environment and position of the robot, only a few applications have been implemented on UAVs. The researches carried out in the LAAS laboratory in France and the Centre for Autonomous Systems in Australia can be highlighted. The first of them has developed an stereo vision system designed for the KARMA blimp [18, 21], where interest point matching and Kalman filtering techniques are used for simultaneous localization and mapping with very good results. However, this approach is not suitable for helicopters, as the baseline of the stereo rig that can be carried is small, and therefore it limits the height at which the UAV can fly. UAV simultaneous localisation and map building with vision using a delta fixed-wing platform is also presented in [19]. Artificial landmarks of known size are used in order to simplify the landmark identification problem. The known size of the landmarks allows to use the cameras as a passive range/bearing/elevation sensor. Preliminary work on the use of vision-based bearing-only SLAM in UAVs is presented in [23]. In [22], vision and IMU are combined for UAV SLAM employing an Unscented Kalman Filter. The feature initialization assumes a flat terrain model, similarly to the present approach. Results in simulation are shown in the paper. In [25], an architecture for multi-vehicle SLAM is studied for its use with UAVs. The paper deals with the issues of data association and communication, and some simulation results are presented.

Visual servoing approaches has been also proposed for direct control of UAVs. The use of an omnidirectional camera for helicopter control has been presented in [17]. The camera is used to maintain the helicopter in the centroid of a set of artificial targets. The processed images are directly used to command the helicopter. The paper shows the feasibility of the procedure, but no actual control is tested. Omnidirectional vision is also used in [12] to estimate the attitude of an UAV. The method detects the horizon line by means of image processing and computes the attitude from its apparent motion. In the work of [27], vision is used to track features of buildings. Image features and GPS measurements are combined together to keep the UAV aligned with the selected features. Control design and stability analysis of image-based controllers for aerial robots are presented in [26]. In [32] recent work on vision-based control of a fixed wing aircraft is presented.

2 Homography-Based Visual Odometry for UAVs

Image homographies will be a basic tool for estimating the motion that an UAV undergoes by using monocular image sequences. A homography can be defined as an invertible application of the projective space $\mathbf{P}^2$ into $\mathbf{P}^2$ that applies lines into lines. Some basic properties of the homographies are the following:

– Any homography can be represented as a linear and invertible transformation in homogeneous coordinates:

$$\begin{bmatrix} \tilde{u} \\ \tilde{v} \\ k \end{bmatrix} = \underbrace{\begin{bmatrix} h_{11} & h_{12} & h_{13} \\ h_{21} & h_{22} & h_{23} \\ h_{31} & h_{32} & h_{33} \end{bmatrix}}_{\mathbf{H}} \begin{bmatrix} u \\ v \\ 1 \end{bmatrix} \tag{1}$$

Inversely, any transformation of this nature can be considered as a homography.
– Given the homogeneous nature of the homography $\mathbf{H}$, it can be multiplied by an arbitrary constant $k \neq 0$ and represent the same transformation. This means that the matrix $\mathbf{H}$ is constrained by eight independent parameters and a scale factor.

Given two views of a scene, the homography model represents the exact transformation of the pixels on the image plane if both views are related by a pure rotation, or if the viewed points lie on a plane. When a UAV flies at relatively high altitude, it is a usual assumption to model the scene as pseudo-planar. The paper will propose a method to extend the applicability of the homography model to non-planar scenes (computing the homography related to a dominant plane on the scene) in order to be able to perform motion estimation at medium or even low UAV altitude.

2.1 Robust Homography Estimation

The algorithm for homography computation is based on a point features matching algorithm, and has been tested and validated with thousands of images captured by different UAVs flying at different altitudes, from 15 to 150 m. This algorithm (including the feature matching approach) was briefly described in [29]. It basically consists of a point-feature tracker that obtains matches between images, and a combination of least median of squares and M-Estimator for outlier rejection and accurate homography estimation from these matches.

However, there are two factors that may reduce the applicability of the technique, mainly when the UAV flies at altitudes of the same order of other elements on the ground (buildings, trees, etc):

– Depending on the frame-rate and the vehicle motion, the overlap between images in the sequence is sometimes reduced. This generates a non-uniform distribution of the features along the images.
– In 3D scenes, the parallax effect will increase, and the planarity assumption will not hold. The result is a dramatic growth of the outliers and even the divergence of the M-Estimator.

They produce different problems when computing the homography. If the matches are not uniformly distributed over the images, an ill-posed system of equations for homography computation will be generated, and there may exist multiple solutions. On the other hand, if the parallax effect is significant, there may exist multiple planes (whose transformation should be described by multiple homographies); the algorithm should try to filter out all features but those lying on the dominant plane of the scene (the ground plane).

In the proposed solution, the first problem is addressed through a *hierarchy* of homographic models (see Fig. 1), in which the complexity of the model to be fitted is decreased whenever the system of equations is ill-constrained, while the second is tackled through outlier rejection techniques.

Therefore, depending on the quality of the available data, the constraints used to compute the homography are different; thus, the accuracy changes as well. An estimation of this accuracy will be given by the covariance matrix of the computed parameters.

A complete homography has 8 *df* (as it is defined up to a scale factor). The degrees of freedom can be reduced by fixing some of the parameters of the 3×3 matrix. The models used are the defined by Hartley in [16]: Euclidean, Affine and Complete Homographic models, which have 4, 6 and 8 *df* respectively (see Fig. 1). The percentage of successful matches obtained by the point tracker is used to have an estimation about the level of the hierarchy where the homography computation should start. These percentage thresholds were obtained empirically by processing hundreds of aerial images. Each level involves the following different steps:

- Complete homography. Least median of squares (LMedS) is used for outlier rejection and a M-Estimator to compute the final result. This model is used if more than the 65% of the matches are successfully tracked.
- Affine homography. If the percentage of success in the tracking step is between 40% and 65%, then the LMedS is not used, given the reduction in the number of matches. A relaxed M-Estimator (soft penalization) is carried out to compute the model.

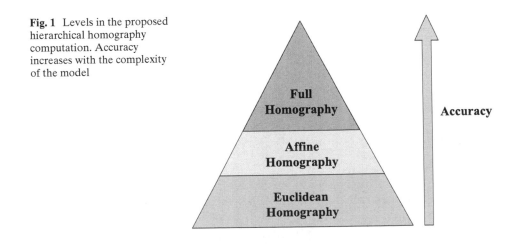

Fig. 1 Levels in the proposed hierarchical homography computation. Accuracy increases with the complexity of the model

– Euclidean homography. If the percentage is below 40%, the set of data is too noisy and small to apply non-linear minimizations. The model is computed using least-squares.

In addition, it is necessary a rule to know when the current level is ill-posed and the algorithm has to decrease the model complexity. The M-Estimator used in the complete and affine computations is used for this purpose. It is considered that the M-Estimator diverge if it reaches the maximum number of iterations and, hence, the level in the hierarchy has to be changed to the next one.

2.2 Geometry of Two Views of the Same Plane

The odometer will extract the camera motion from the image motion modeled by the estimated homography between two consecutive views. If we consider the position and orientation of two cameras in the world coordinate frame, as shown in Fig. 2, it can be seen that the two projections $\mathbf{m_1} \in \mathbb{R}^2$ and $\mathbf{m_2} \subset \mathbb{R}^2$ of a fixed point $\mathbf{P} \in \mathbb{R}^3$ belonging to a plane Π are related by:

$$\tilde{\mathbf{m}}_2 = \underbrace{\mathbf{A}_2\mathbf{R}_{12}\left(\mathbf{I} - \frac{\mathbf{t}_2\mathbf{n}_1^T}{d_1}\right)\mathbf{A}_1^{-1}}_{\mathbf{H}_{12}} \tilde{\mathbf{m}}_1 \qquad (2)$$

where $\mathbf{R}_{12}$ is the rotation matrix that transforms a vector expressed in the coordinate frame of camera one into the coordinate frame of camera two, $\mathbf{t}_2$ is the translation of camera two with respect to camera one expressed in the coordinate frame of camera one, the Euclidean distance from the camera one to the plane Π is d_1 and the normal of the plane Π (in the first camera coordinate frame) is given by the unitary 3-D vector $\mathbf{n}_1$ (see Fig. 2).

Fig. 2 Geometry of two views of the same plane

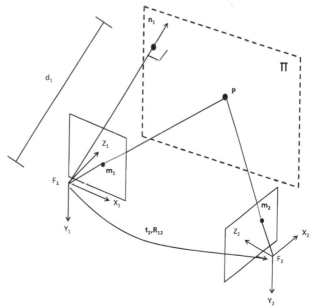

For this particular case, the transformation between the features $\mathbf{m}_1$ and $\mathbf{m}_2$ is a plane-to-plane homography, so $\tilde{\mathbf{m}}_2 = \mathbf{H}_{12}\tilde{\mathbf{m}}_1$. This homography is completely defined by the calibration matrices $\mathbf{A}_1$ and $\mathbf{A}_2$, the relative position of the cameras and the structure of the scene (the normal and distance of the plane). The problem can be reformulated as a single camera whose position and orientation change through time. In this case the calibration matrix is the same for both views, so $\mathbf{A}_1 = \mathbf{A}_2$.

Then, for the calibrated case, the relative position (rotation and translation) between the cameras and the plane normal can be obtained if the homography that relates two views of the same plane is known, for instance by obtaining a set of matches between the images, as described in the previous section. Moreover, it will be shown how to obtain an estimation of the covariance matrix for all these parameters.

2.3 Motion Estimation from Homographies

A solution based on the singular value decomposition (SVD) of the homography will be used. Consider a single camera that moves through time, the homography $\mathbf{H}_{12}$ that relates the first and the second view of the same planar scene and the camera calibration matrix $\mathbf{A}_1$. According to Eq. 2, the *calibrated homography* is defined as:

$$\mathbf{H}_{12}^c = \mathbf{A}_1^{-1}\mathbf{H}_{12}\mathbf{A}_1 = \mathbf{R}_{12}\left(\mathbf{I} - \frac{\mathbf{t}_2\mathbf{n}_1^T}{d_1}\right) \tag{3}$$

The elements can be extracted from the singular value decomposition (SVD) of the homography $\mathbf{H}_{12}^c = \mathbf{U}\mathbf{D}\mathbf{V}^T$, where $\mathbf{D} = diag(\lambda_1, \lambda_2, \lambda_3)$ stores the singular values. Once $\mathbf{U}$, $\mathbf{V}$ and $\mathbf{D}$ have been conveniently ordered such us $\lambda_1 > \lambda_2 > \lambda_3$, the singular values can be used to distinguish three types of movements carried out by the camera [39]:

- The three singular values of $\mathbf{H}_{12}^c$ are equal, so $\lambda_1 = \lambda_2 = \lambda_3$. It occurs when the motion consist of rotation around an axis through the origin only, i.e., $\mathbf{t}_2 = \mathbf{0}$. The rotation matrix is unique, but there is not sufficient information to estimate the plane normal $\mathbf{n}_1$.
- The multiplicity of the singular values of $\mathbf{H}_{12}^c$ is two, for example $\lambda_1 = \lambda_2 \neq \lambda_3$. Then, the solution for motion and geometrical parameters is unique up to a common scale factor for the translation parameters. In this case, the camera translation is parallel to the normal plane.
- The three singular values of $\mathbf{H}_{12}^c$ are different, i.e., $\lambda_1 \neq \lambda_2 \neq \lambda_3$. In this case two possible solutions for rotation, translation and plane normal exist and can be computed.

The presence on noise in both feature tracking and homography estimation always leads to different singular values for $\mathbf{H}_{12}^c$ and the third of the previous cases becomes the dominant in real conditions. Rotation, translation and normal to the plane is then given by the following expressions [39]:

$$\mathbf{R}_2 = \mathbf{U} \begin{bmatrix} \alpha & 0 & \beta \\ 0 & 1 & 0 \\ -s\beta & 0 & s\alpha \end{bmatrix} \mathbf{V}^T$$

$$\mathbf{t}_2 = \frac{1}{w} \left(-\beta \mathbf{u}_1 + \left(\frac{\lambda_3}{\lambda_2} - s\alpha \right) \mathbf{u}_3 \right)$$

$$\mathbf{n}_1 = w(\delta \mathbf{v}_1 + \mathbf{v}_3) \tag{4}$$

where:

$$\delta = \pm \sqrt{\frac{\lambda_1^2 - \lambda_2^2}{\lambda_2^2 - \lambda_3^2}}$$

$$\alpha = \frac{\lambda_1 + s\lambda_3 \delta^2}{\lambda_2 (1 + \delta^2)}$$

$$\beta = \pm \sqrt{1 - \alpha^2}$$

$$s = det(\mathbf{U}) det(\mathbf{V})$$

and ω is a scale factor. We set that scale factor so that $\|\mathbf{n}_1\| = 1$. Each solution must accomplish that $sgn(\beta) = -sgn(\delta)$. For this case, Triggs algorithm [38] allows a systematic and robust estimation. This method has been implemented and tested in the experiments presented in this paper with very good results.

From Eq. 3 it can be seen that (as $\|\mathbf{n}_1\| = 1$) only the product $\frac{\|\mathbf{t}_2\|}{d_1}$ can be recovered. The scale can be solved, then, if the distance d_1 of camera 1 to the reference plane is known. If the reference plane is the ground plane, as it would be the case in the experiments, a barometric sensor or height sensor can be used to estimate this initial distance. Also, a range sensor can be used. In this paper, we will consider that this height is estimated for the first frame by one of these methods.

2.4 Correct Solution Disambiguation

Apart from the scale factor, two possible solutions $\{\mathbf{R}_2^1, \mathbf{t}_2^1, \mathbf{n}_1^1\}$ and $\{\mathbf{R}_2^2, \mathbf{t}_2^2, \mathbf{n}_1^2\}$ will be obtained. Given a third view and its homography with respect to the first frame $\mathbf{H}_{13}$, it is possible to recover an unique solution, as the estimated normal of the reference plane in the first camera coordinate frame, $\mathbf{n}_1$, should be the same.

A method to detect the correct solution is proposed. If a sequence of images is used, the set of possible normals is represented by:

$$S_n = \{\mathbf{n}_{12}^1, \mathbf{n}_{12}^2, \mathbf{n}_{13}^1, \mathbf{n}_{13}^2, \mathbf{n}_{14}^1, \mathbf{n}_{14}^2, ...\} \tag{5}$$

where the superindex denotes the two possible normal solutions and the subindex $1j$ denotes the normal $\mathbf{n}_1$ estimated using image j in the sequence.

If $\mathbf{n}_{12}^1$ and $\mathbf{n}_{12}^2$ were correct, there would be two set of solutions, S_{n^1} and S_{n^2}. The uniqueness of the normal leads to the following constraints:

$$\left\| \mathbf{n}_{12}^1 - \mathbf{n}_{1j}^i \right\| \leq \epsilon_1 \ \forall \mathbf{n}_{1j}^i \in S_{n^1} \tag{6}$$

$$\left\| \mathbf{n}_{12}^2 - \mathbf{n}_{1j}^i \right\| \leq \epsilon_2 \ \forall \mathbf{n}_{1j}^i \in S_{n^2} \tag{7}$$

where ϵ_1 and ϵ_2 are the minimal values that guarantee an unique solution for Eqs. 6 and 7 respectively. The pairs $\{S_{n^1}, \epsilon_1\}$ and $\{S_{n^2}, \epsilon_2\}$ are computed separately by means of the following iterative algorithm:

1. The distance among $\mathbf{n}_{12}^i$ and the rest of normals of S_n is computed.
2. ϵ_i is set to an initial value.
3. For the current value ϵ_i, check if there exist an unique solution.
4. If no solution is found, increase the value of ϵ_i and try again with the step 3. If multiple solutions were found decrease ϵ_i and try again with step 3. If an unique solution was found, then finish.

The algorithm is applied to $i = 1$ and $i = 2$ and the correct solution is then chosen between both options as the one that achieves the minimum ϵ.

2.5 An Estimation of the Uncertainties

An important issue with odometric measurements is to obtain a correct estimation of the associated drift. The idea is to estimate the uncertainties on the estimated rotation, translation and plane normal from the covariance matrix associated to the homography, which can be computed from the estimated errors on the point matches [6].

The proposed method computes the Jacobian of the complete process to obtain a first order approximation of rotation, translation and plane normal error covariance matrix. Once the calibrated homography has been decomposed into its singular values, the computation of the camera motion is straightforward, so this section will focus in the computation of the Jacobian associated to the singular value decomposition process.

Thus, given the SVD decomposition of the calibrated homography $\mathbf{H}_{12}^c$:

$$\mathbf{H}_{12}^c = \begin{bmatrix} h_{11} & h_{12} & h_{13} \\ h_{21} & h_{22} & h_{23} \\ h_{31} & h_{32} & h_{33} \end{bmatrix} = \mathbf{U}\mathbf{D}\mathbf{V}^T = \sum_{i=1}^{3} \left(\lambda_i \mathbf{u}_i \mathbf{v}_i^T \right) \tag{8}$$

The goal is to compute $\frac{\partial \mathbf{U}}{\partial h_{ij}}$, $\frac{\partial \mathbf{V}}{\partial h_{ij}}$ and $\frac{\partial \mathbf{D}}{\partial h_{ij}}$ for all h_{ij} in $\mathbf{H}_{12}^c$. This Jacobian can be easily computed through the robust method proposed by Papadopoulo and Lourakis in [31].

Taking the derivative of Eq. 8 with respect to h_{ij} yields the following expression:

$$\frac{\partial \mathbf{H}_{12}^c}{\partial h_{ij}} = \frac{\partial \mathbf{U}}{\partial h_{ij}}\mathbf{D}\mathbf{V}^T + \mathbf{U}\frac{\partial \mathbf{D}}{\partial h_{ij}}\mathbf{V}^T + \mathbf{U}\mathbf{D}\frac{\partial \mathbf{V}^T}{\partial h_{ij}} \tag{9}$$

Clearly, $\forall (k, l) \neq (i, j)$, $\frac{\partial h_{kl}}{\partial h_{ij}} = 0$ while $\frac{\partial h_{ij}}{\partial h_{ij}} = 1$. Since $\mathbf{U}$ is an orthogonal matrix:

$$\mathbf{U}^T \mathbf{U} = \mathbf{I} \Rightarrow \frac{\partial \mathbf{U}^T}{\partial h_{ij}} \mathbf{U} + \mathbf{U}^T \frac{\partial \mathbf{U}}{\partial h_{ij}} = \Omega_U^{ij}{}^T + \Omega_U^{ij} = 0 \tag{10}$$

where Ω_U^{ij} is defined by

$$\Omega_U^{ij} = \mathbf{U}^T \frac{\partial \mathbf{U}}{\partial h_{ij}} \tag{11}$$

It is clear that Ω_U^{ij} is an antisymmetric matrix. Similarly, an antisymmetric matrix Ω_V^{ij} can be defined for $\mathbf{V}$ as:

$$\Omega_V^{ij} = \frac{\partial \mathbf{V}^T}{\partial h_{ij}} \mathbf{V} \tag{12}$$

By multiplying Eq. 9 by $\mathbf{U}^T$ and $\mathbf{V}$ from left and right respectively, and using Eqs. 11 and 12, the following relation is obtained:

$$\mathbf{U}^T \frac{\partial \mathbf{H}_{12}^u}{\partial h_{ij}} \mathbf{V} - \Omega_U^{ij} \mathbf{D} + \frac{\partial \mathbf{D}}{\partial h_{ij}} + \mathbf{D} \Omega_V^{ij} \tag{13}$$

Since Ω_U^{ij} and Ω_V^{ij} are antisymmetric matrices, all their diagonal elements are equal to zero. Recalling that $\mathbf{D}$ is a diagonal matrix, it is easy to see that the diagonal elements of $\Omega_U^{ij} \mathbf{D}$ and $\mathbf{D} \Omega_V^{ij}$ are also zero. Thus:

$$\frac{\partial \lambda_k}{\partial h_{ij}} = u_{ik} v_{jk} \tag{14}$$

Taking into account the antisymmetric property, the elements of the matrices Ω_U^{ij} and Ω_V^{ij} can be computed by solving a set of 2×2 linear systems, which are derived from the off-diagonal elements of the matrices in Eq. 13:

$$\left. \begin{array}{l} d_l \Omega_{U\,kl}^{ij} + d_k \Omega_{V\,kl}^{ij} = u_{ik} v_{jl} \\ d_k \Omega_{U\,kl}^{ij} + d_l \Omega_{V\,kl}^{ij} = -u_{il} v_{jk} \end{array} \right\} \tag{15}$$

where the index ranges are $k = 1 \ldots 3$ and $l = i + 1 \ldots 2$. Note that, since the d_k are positive numbers, this system has a unique solution provided that $d_k \neq d_l$. Assuming for now that $\forall (k, l)$, $d_k \neq d_l$, the 3 parameters defining the non-zero elements of Ω_U^{ij} and Ω_V^{ij} can be easily recovered by solving the 3 corresponding 2×2 linear systems.

Once Ω_U^{ij} and Ω_V^{ij} have been computed, the partial derivatives are obtained as follows:

$$\frac{\partial \mathbf{U}}{\partial h_{ij}} = \mathbf{U} \Omega_U^{ij} \tag{16}$$

$$\frac{\partial \mathbf{V}}{\partial h_{ij}} = -\mathbf{V} \Omega_V^{ij} \tag{17}$$

Taking into account the Eqs. 14, 16 and 17 and the covariance matrix corresponding to the homography it is possible to compute the covariance matrix associated to $\mathbf{U}$, $\mathbf{V}$ and $\mathbf{D}$. Further details and demonstrations can be found in [31]. Finally, the

Jacobians of the equations used to extract the rotation, translation and normal, given by Eq. 4, are easily computed and combined with these covariances to estimate the final motion covariances.

2.6 Experimental Results

This section shows some experimental results in which the homography-based visual odometer is applied to monocular image sequences gathered by real UAVs.

The first experiment was conducted with the HERO helicopter (see Fig. 3). HERO is an aerial robotic platform designed for research on UAV control, navigation and perception. It has been developed by the "Robotics, Vision and Control Research Group" at the University of Seville during the CROMAT project, funded by the Spanish Government. HERO is equipped with accurate sensors to measure position and orientation, cameras and a PC-104 to allow processing on board. A DSP is used as data acquisition system and low level controller (position and orientation); the PC-104 runs the rest of tasks such as perception, communications or navigation. All the data gathered by the DSP are exported to the PC-104 through a serial line and published for the rest of the processes.

All the sensor data have been logged together with the images in order to avoid inconsistency among different sensor data. The position is estimated with a Novatel DGPS with 2 cm accuracy and updated at 5 Hz, while an inertial measurement unit (IMU) provides the orientation at 50 Hz, with accuracy of 0.5 degrees. In the experiment, the camera was oriented forty-five degrees with respect to the helicopter horizontal.

The visual odometer algorithm (feature tracking, robust homography computation and homography decomposition) has been programmed in C++ code and runs at 10 Hz with 320 × 240 images. The experiment image sequence is composed by 650 samples, or approximately 65 s of flight. A sharp movement is made around sample 400.

The DGPS measurements are used to validate the results. Along the flight, good GPS coverage was available at all time. It is important to notice that the

Fig. 3 HERO helicopter

odometry is computed taking into account the estimated translation and rotation, so it accumulates both errors. The estimated position by using the visual odometer is shown in Fig. 4. The figure presents the DGPS position estimation and the errors associated to the odometer. It can be seen how the errors grow with the image sample index. The errors corresponding to each estimation are added to the previous ones

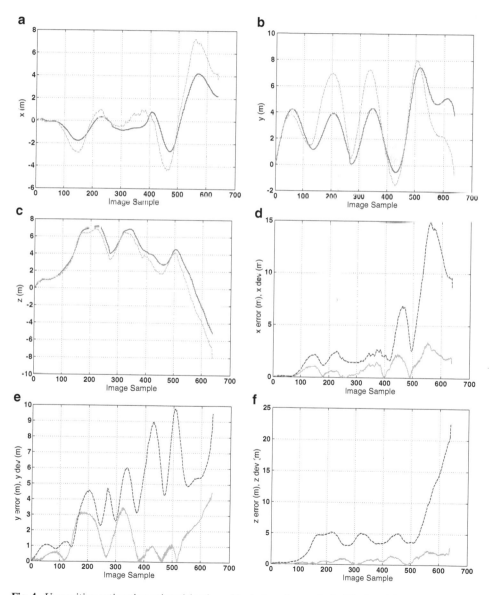

Fig. 4 *Up* position estimation using vision based technique (*green dashed line*) and DGPS estimation (*red solid line*). *Down* error of the vision based odometry (*green solid line*) and estimated standard deviation (*blue dashed line*)

and make the position estimation diverge through time. Moreover, it can be seen how the estimation of the standard deviation is coherent with the evolution of the error (which is very important for further steps).

Figure 5 shows the evolution of the estimated orientation by using the odometer and the on-board IMU. The orientation has been represented in the classic

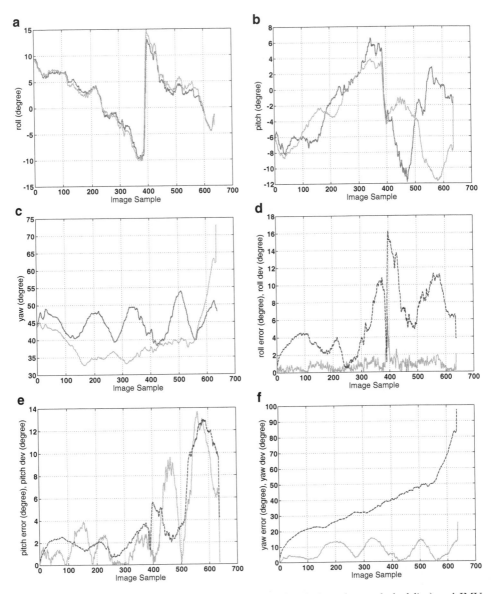

Fig. 5 *Top* estimated orientation by using vision based technique (*green dashed line*) and IMU estimation (*red solid line*). The orientation is represented in roll/pitch/yaw. *Bottom* errors in the vision based estimation (*green solid line*) and estimated standard deviation (*blue dashed line*)

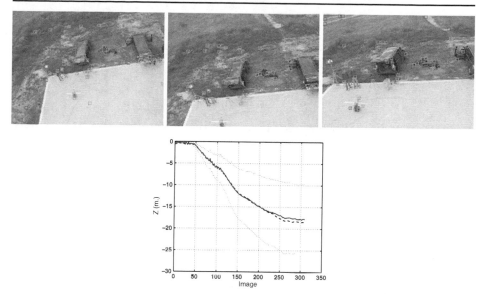

Fig. 6 Three images of the landing sequence and the estimated height computed by the visual odometer (*solid*) and DGPS (*dashed*). The average frame rate is 7 Hz

roll/pitch/yaw convention (Euler XYZ). It can be seen that the errors in the estimated orientation are small except for the pitch angle. The standard deviation is in general overall consistent.

Results have been also obtained with data gathered during an autonomous landing[1] by the autonomous helicopter Marvin, developed by the Technical University of Berlin [33]. Figure 6 shows three frames of the landing sequence with the obtained matches. It should be pointed out that there are no artificial landmarks for the matching process. Also, in this experiment, the concrete landing platform lacks of structure, which can pose difficulties for the matching procedure. Moreover, along the descent, the pan and tilt unit was moving the camera. Figure 6 shows the estimated translation compared with DGPS, along with the estimated errors. The results are very accurate, although the technique tends to overestimate the uncertainty.

Thus, the experimental results show that the visual odometer can be used to estimate the motion of the UAV; moreover, the estimated errors are consistent. It is important to highlight that all experiments were carried out by using natural landmarks automatically selected by the feature tracking algorithm, without the help of visual beacons.

3 Application of Homography-Based Odometry to the SLAM Problem

A SLAM-based technique is proposed to compensate the accumulative error intrinsic to odometry and to solve the localization problem. SLAM employing monocular

---

[1]The autonomous landing was done based on DGPS and ultrasonic sensors.

imagery is a particular case of the SLAM problem, called bearing-only SLAM or boSLAM, in which bearing only sensors are used, a camera in this case. boSLAM is a partially observable problem [41], as the depth of the landmarks cannot be directly estimated. This entails a difficult landmark initialization problem which has been tackled with two basic approaches: delayed and un-delayed initialization. In the delayed initialization case, landmarks are not included in the SLAM system in the first observation, but when the angular baseline in between observations has grown large enough to ensure a good triangulation. This method has the advantage of using well conditioned landmarks, but the SLAM system cannot take advantage of the landmark until its localization is well conditioned. Several approaches have been proposed in this area such as [10] where a Particle Filter is used to initialize the landmark depth, or [11], where non-linear bundle adjustment over a set of observations is used to initialize the landmarks.

On the other hand, un-delayed approaches introduce the landmark in the SLAM system with the first observation, but some considerations have to be taken into account due to the fact that the landmarks are usually bad conditioned in depth, and then divergence problems may appear in the SLAM filter. Most existing approaches are based on multiple hypotheses, as in [20], where a Gaussian Mixture is used for landmark initialization in a Kalman Filter. Recent research [28] proposes the inverse depth parametrization in a single-hypothesis approach for landmark initialization.

The technique presented in this Section is based on a classical EKF that simultaneously estimates the pose of the robot (6 df) and a map of point features, as in [2, 3, 14, 24]. The main contributions is a new undelayed feature initialization that takes advantage of the scene normal plane estimation computed in the Homography-based odometry. Indeed, the technique cannot be considered as boSLAM because information from a range sensor is used, combined with the normal vector to the plane, to initialize the landmark depth.

The use of the estimated rotation and translation provided by the odometer as the main motion hypothesis in the prediction stage of the EKF is another contribution made by this approach. Complex non-linear models are normally used to estimate vehicle dynamics, due to the lack of odometers in UAVs. This leads to poor prediction hypotheses, in terms of accuracy, and then a significant reduction of the filter efficiency. In [19] a solution based on merging model-based estimation and inertial measurements from local sensors (IMUs) is proposed, resulting in an accuracy growth. The integration of the IMU is also considered here in order to improve the position estimation. Next paragraphs describe the structure and implementation of this filter.

3.1 The State Vector

The robot pose $\mathbf{p}_t$ is composed by the position and orientation of the vehicle at time t in the World Frame (see Section 3.4), so:

$$\mathbf{p}_t = [\mathbf{t}_t, \mathbf{q}_t]^T = [x, y, z, q_x, q_y, q_z, q_w]^T \tag{18}$$

where $\mathbf{t}_t$ expresses the position at time t of the UAV in the world coordinate frame, and $\mathbf{q}_t$ is the unitary quaternion that aligns the robot to the world reference frame at time t. Using quaternions increases (in one) the number of parameters for the orientation with respect to Euler angles, but simplifies the algebra and hence, the error propagation. However, the quaternion normalization has to be taken into account after the prediction and update stages.

Landmarks will be represented by their 3D cartesian position in the World Frame $\mathbf{y}_n$. Thus, the state vector $\mathbf{x}_t$ is composed by the robot pose $\mathbf{p}_t$ and the set of current landmarks $\{\mathbf{y}_1, ..., \mathbf{y}_n\}$ so:

$$\mathbf{x}_t = \left[\mathbf{p}_t^T, \mathbf{y}_1^T, ..., \mathbf{y}_n^T\right]^T \tag{19}$$

3.2 Prediction Stage

Given the pose at time $t - 1$, the odometer provides the translation with respect to the previous position (expressed in the $t - 1$ frame) and the rotation that transforms the previous orientation into the new one (expressed in the t frame). Taking into account the quaternions algebra, the state vector at time t can be computed as:

$$\mathbf{t}_t = \mathbf{t}_{t-1} + \mathbf{q}_{t-1} \otimes \mathbf{t}_u \otimes \mathbf{q}_{t-1}^{-1} \tag{20}$$

$$\mathbf{q}_t = \mathbf{q}_u^{-1} \otimes \mathbf{q}_{t-1} \tag{21}$$

where $\mathbf{t}_u$ and $\mathbf{q}_u$ represent the estimated translation and rotation from the odometer, and $\otimes$ denotes quaternion multiplication. Notice that prediction does not affect the landmark position because they are assumed to be motionless.

Computing the odometry requires to carry out the image processing between consecutive images detailed in Section 2: feature tracking, homography estimation and, finally, odometry. The estimated translation and rotation covariance matrices are used to compute the process noise covariance matrix.

3.3 Updating Stage

From the whole set of features provided by the feature tracking algorithm used in the prediction stage, a small subset is selected to act as landmarks. The features associated to the landmarks are taken apart and not used for the homography estimation in order to eliminate correlations among prediction and updating. Thus, the number of landmarks must be a compromise between the performance of the EKF and the performance of the homography estimation (and thus, the odometry estimation). In addition, the computational requirements of the full SLAM approach has to be considered.

Experimental results allowed the authors to properly tune the number of landmarks and features used in the approach. A set of one hundred features are tracked from one image to another, and a subset of ten/fifteen well distributed and stable features are used as landmarks. Therefore, for each new image, the new position of this subset of features will be given by the feature tracking algorithm; this information will be used as measurement at time t, $\mathbf{z}_t$.

If the prediction stage was correct, the projection of each landmark into the camera would fit with the estimated position of the feature given by the tracking algorithm. If the landmark $\mathbf{y}_n$ corresponds to the image feature $\mathbf{m}_n = [u, v]$, following the camera projection model (Fig. 7):

$$\tilde{\mathbf{m}}_n = \mathbf{A}\left(\mathbf{q}_t^{-1} \otimes (\mathbf{y}_n - \mathbf{t}_t) \otimes \mathbf{q}_t\right) \tag{22}$$

Fig. 7 Projection of a landmark into the camera. The landmark is represented by a *black dot*, the translation of the camera focal (F) in the world frame ($\mathcal{W}$) is $\mathbf{t}_t$, the back-projection of the feature $\mathbf{m}_n$ is $\tilde{\mathbf{m}}_n$ and the position of the landmark in the world frame is $\mathbf{y}_n$

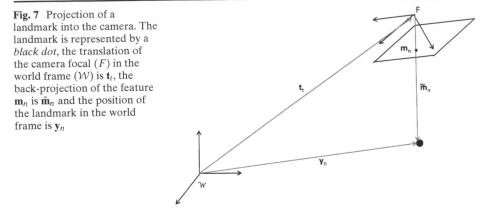

where $\mathbf{A}$ is the camera calibration matrix and $\tilde{\mathbf{m}}_n = [\tilde{u}, \tilde{v}, h]$, so the feature position is computed as $\mathbf{m}_n = [\tilde{u}/h, \tilde{v}/h]$. This measurement equation is applied to all the features correctly tracked from the previous image to the current one. The data association problem is solved by means of the feature matching algorithm.

In order to bound the computational cost needed for the SLAM approach, landmarks are not stored indefinitely in the EKF filter. Instead, they are maintained for a short period of time in the filter just to avoid transient occlusions, later they are automatically marginalized out from the filter and a new feature, provided by the tracker, initialized. If the corresponding landmarks are well conditioned, the measurement equation constraints the current position and orientation of the UAV.

3.4 Filter and Landmarks Initialization

The filter state vector will be initialized to a given position and orientation. This information can be provided by external devices such as GPS and IMU, and the process covariance matrix to the corresponding error information. The position can be also initialized to zero, so the first position is assumed as the origin and the corresponding covariances are zero too. This initial position defines the World Reference Frame where the landmarks and UAV pose are expressed.

In the following, a more sophisticated method for landmark initialization is proposed. When a new image feature is selected for being a landmark in the filter, it is necessary to compute its real position in the World frame. Due to the bearing only nature of the camera, the back-projection of the feature is given by a ray defined by the camera focal point and the image of the landmark. The proposed technique takes advantage of knowing the normal to the scene plane and the distance from the UAV to the ground at a given time. With this information the ground can be locally approximated by a plane and the landmark position as the intersection of the back-projection ray with this plane, as shown in Fig. 8.

If the World frame is aligned with the camera frame, the back-projection of the feature $\mathbf{m}_n = [u, v]$ will be the ray $\mathbf{r}$ defined by:

$$\mathbf{r} : \mathbf{A}^{-1}\tilde{\mathbf{m}}_n \tag{23}$$

Fig. 8 Landmark
initialization representation

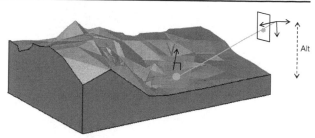

where $\mathbf{A}$ is the camera calibration matrix and $\tilde{\mathbf{m}}_n = [h\mathbf{m}_n, h]$. In addition, the odometer provides an estimation of the normal to the scene plane at time t denoted as $\mathbf{n}_t$. Given the distance to the plane d_t, the plane Π is defined as:

$$\Pi : d_t \quad \mathbf{n}_t^T \begin{bmatrix} x \\ y \\ z \end{bmatrix} \tag{24}$$

Then, the landmark position will be computed as the intersection of the ray $\mathbf{r}$ with the plane Π. If Eqs. 23 and 24 are merged, the value of λ can be easily computed as:

$$h = \left(\mathbf{n}_t^T \mathbf{A}^{-1} \tilde{\mathbf{m}}_n\right)^{-1} d_t \tag{25}$$

and the landmark can be computed as:

$$\mathbf{y}_n = \left(\mathbf{n}_t^T \mathbf{A}^{-1} \tilde{\mathbf{m}}_n\right)^{-1} d_t \mathbf{A}^{-1} \tilde{\mathbf{m}}_n \tag{26}$$

But this landmark is expressed in the camera coordinate frame. The UAV current position $\mathbf{d}_t$ and orientation $\mathbf{q}_t$ are finally used to express the landmark in the World frame:

$$\mathbf{y}_n = \mathbf{t}_t + \mathbf{q}_t \otimes \left(\left(\mathbf{n}_t^T \mathbf{A}^{-1} \tilde{\mathbf{m}}_n\right)^{-1} d_t \mathbf{A}^{-1} \tilde{\mathbf{m}}_n\right) \otimes \mathbf{q}_t^{-1} \tag{27}$$

There is a strong dependence of this approach on the planarity of the scene. The more planar the scene is, the better the plane approximation, leading to smaller noise in the plane normal estimation, and thus, to a better initialization.

Nevertheless, the back-projection procedure is still non-linear, and therefore, the Gaussian approximation for the errors has to be carefully considered. If the relative orientation of the ray $\mathbf{r}$ associated to a feature is near parallel with respect to the plane, the errors on the estimation can be high, and a Gaussian distribution will not approximate the error shape adequately. Then, only those landmarks for which the relative orientation of the ray and the plane is higher than 30 degrees will be considered in the initialization process.

3.5 Experimental Results on Homography-Based SLAM

To test the proposed approach, experiments with the HERO helicopter were carried out. The image sequence was gathered at 15 m of altitude with respect to the ground and with the camera pointed 45 degrees with respect to the helicopter horizontal.

It is important to remark that no close-loop was carried out during the experiment, although there are some loops present in the UAV trajectory, this subject is out of the scope of this research work. Therefore, the result can be improved if a reliable data association algorithm is used for detecting and associating landmarks in the filter. The complete size of the trajectory is about 90 m long.

IMU information is used to express the results in the same frame than DGPS measurements. The results of the experiment are shown in Fig. 9, where the estimation

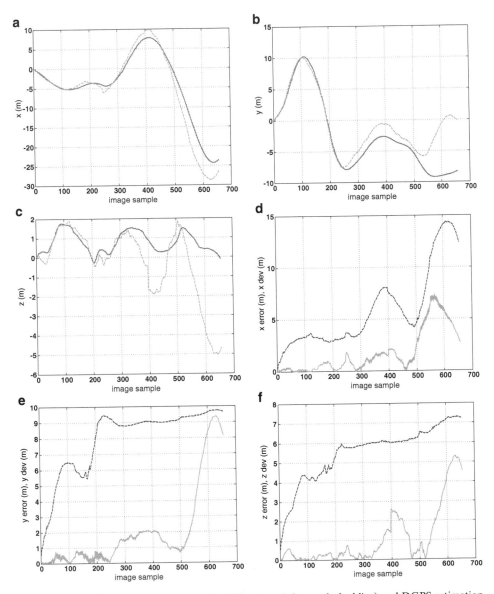

Fig. 9 *Up* position estimation using the SLAM approach (*green dashed line*) and DGPS estimation (bi). *Down* error of the SLAM approach (*green solid line*) and estimated standard deviation (*blue dashed line*)

Fig. 10 XY position estimation using the SLAM approach (*green dashed line*) and DGPS estimation (*red solid line*)

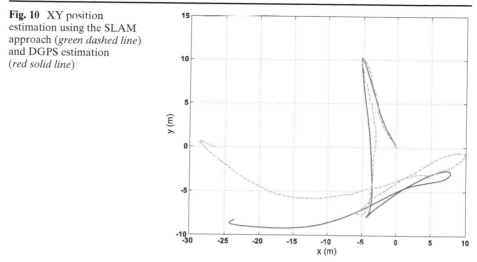

in each axis and the errors (with respect to DGPS outputs) are plotted. It can be seen how the uncertainty estimation is coherent with the measured errors. However, the position slowly diverges through time due to the absence of large loop closing. The instant orientation is not plotted because it is inherently taken into account in the computation of position. More details are shown in Fig. 10, where the XY DGPS trajectory is plotted together with the XY estimation.

3.6 Experimental Results Including an Inertial Measurement Unit

The errors shown in Fig. 9 are partially generated by a drift in the estimation of the UAV orientation. If the measurements of an inertial measurement unit (IMU) are incorporated into the SLAM approach, the errors introduced by the orientation estimation can be reset, and then the localization could be improved.

Fig. 11 XY position estimation using the SLAM approach with IMU corrections (*green dashed line*) and DGPS estimation (*red solid line*)

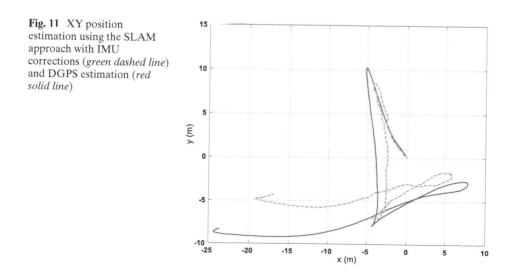

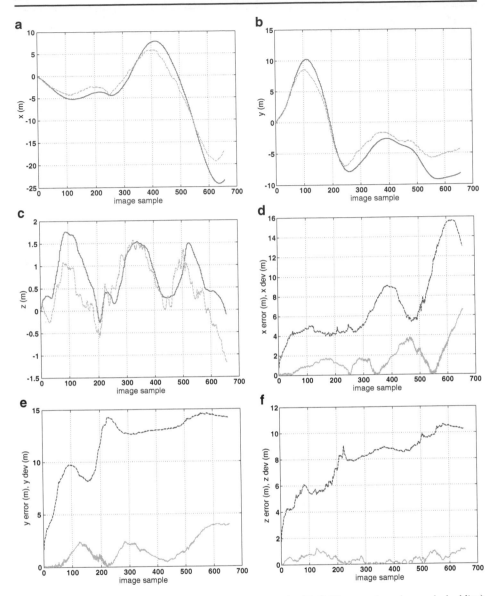

Fig. 12 *Up* position estimation using the SLAM approach with IMU corrections (*green dashed line*) and DGPS estimation (*red solid line*). *Down* error of the SLAM approach with IMU corrections (*green solid line*) and estimated standard deviation (*blue dashed line*)

The proposed SLAM approach can be easily adapted to include the IMU information by integrating its measurement in the prediction stage of the EKF. The IMU provides the complete orientation, so there is no error integration and it is bounded by the accuracy of the device.

This approach has been tested with the same data set. The XY estimation is plotted in Fig. 11. Figure 12 shows the estimation compared to the DGPS measurement. It can be seen that the errors in Z and Y are significantly smaller while in X are slightly smaller with respect to the approach without considering the IMU.

4 Conclusions

The paper presents contributions to the vision-based navigation of aerial vehicles. It is proposed a visual odometry system for UAVs based on monocular imagery. Homographic models and homography decomposition are used to extract the real camera motion and the normal vector to the scene plane. A range sensor is used to obtain the scale factor of the motion. The paper shows the feasibility of the approach through experimental results with real UAVs.

An important aspect of the proposed odometry approach is the use of natural landmarks instead of beacons or visual references with known positions. A general-purpose feature tracking is used for this purpose. Although natural landmarks increase the applicability of the proposed techniques, they also increase the complexity of the problem to be solved. In fact, outlier rejection and robust homography estimation are required.

The paper also proposes a localization technique based on monocular vision. An Extended Kalman filter-based SLAM is successfully used to compute the localization and mapping. Two basic contributions to SLAM with UAVs are proposed. First, the use of a vision based odometry as main motion hypothesis for the prediction stage of the Kalman filter and, second, a new landmark initialization technique that exploits the benefits of estimating the normal to the scene plane. Both techniques are implemented and validated with a real UAV.

Although no large loops are closed in the experiments, the estimated position and covariance are coherent, so the result could be improved if a reliable data association algorithm is used for detecting and associating landmarks in the filter.

Future developments will consider different features with better invariance characteristics in order to close loops. It should be pointed out that the method can be applied to piece-wise planar scenes, like in urban scenarios

Acknowledgements The authors would like to thank the Technical University of Berlin for providing images to validate the homography-based odometry approach. In addition, the authors thank the support of Victor Vega, Fran Real and Ivan Maza during the experiments with HERO helicopter.

References

1. Amidi, O., Kanade, T., Fujita, K.: A visual odometer for autonomous helicopter fight. In: Proceedings of the Fifth International Conference on Intelligent Autonomous Systems (IAS-5), June 1998
2. Betge-Brezetz, S., Hebert, P., Chatila, R., Devy, M.: Uncertain map making in natural environments. In: Proceedings of the IEEE International Conference on Robotics and Automation, vol. 2, pp. 1048–1053, April 1996
3. Betke, M., Gurvits, L.: Mobile robot localization using landmarks. IEEE Trans. Robot. Autom. **13**, 251–263 (1997)

4. Byrne, J., Cosgrove, M., Mehra, R.: Stereo based obstacle detection for an unmanned air vehicle. In: Proceedings 2006 IEEE International Conference on Robotics and Automation, pp. 2830–2835, May 2006
5. Caballero, F., Merino, L., Ferruz, J., Ollero, A.: A visual odometer without 3D reconstruction for aerial vehicles. applications to building inspection. In: Proceedings of the International Conference on Robotics and Automation, pp. 4684–4689. IEEE, April 2005
6. Caballero, F., Merino, L., Ferruz, J., Ollero, A.: Improving vision-based planar motion estimation for unmanned aerial vehicles through online mosaicing. In: Proceedings of the International Conference on Robotics and Automation, pp. 2860–2865. IEEE, May 2006
7. Caballero, F., Merino, L., Ferruz, J., Ollero, A.: Homography based Kalman filter for mosaic building. applications to UAV position estimation. In: IEEE International Conference on Robotics and Automation, pp. 2004–2009, April 2007
8. Conte, G., Doherty, P.: An integrated UAV navigation system based on aerial image matching. In: Proceedings of the IEEE Aerospace Conference, pp. 1–10 (2008)
9. Corke, P.I., Sikka, P., Roberts, J.M.: Height estimation for an autonomous helicopter. In: Proceedings of ISER, pp. 101–110 (2000)
10. Davison, A.: Real-time simultaneous localisation and mapping with a single camera. In: IEEE International Conference on Computer Vision, pp. 1403–1410, October 2003
11. Deans, M., Hebert, M.: Experimental comparison of techniques for localization and mapping using a bearings only sensor. In: Proceedings of the Seventh International Symposium on Experimental Robotics, December 2000
12. Demonceaux, C., Vasseur, P., Pegard, C.: Omnidirectional vision on UAV for attitude computation. In: Proceedings 2006 IEEE International Conference on Robotics and Automation, pp. 2842–2847, May 2006
13. Dickmanns, E.D., Schell, F.R.: Autonomous landing of airplanes using dynamic machine vision. In: Proc. of the IEEE Workshop Applications of Computer Vision, pp. 172–179, December 1992
14. Feder, H.J.S., Leonard, J.J., Smith, C.M.: Adaptive mobile robot navigation and mapping. Int. J. Rob. Res. **18**(7), 650–668 (1999) July
15. Garcia-Pardo, P.J., Sukhatme, G.S., Montgomery, J.F.: Towards vision-based safe landing for an autonomous helicopter. Robot. Auton. Syst. **38**(1), 19–29 (2001)
16. Hartley, R.I., Zisserman, A.: Multiple View Geometry in Computer Vision, 2nd edn. Cambridge University Press (2004)
17. Hrabar, S., Sukhatme, G.S.: Omnidirectional vision for an autonomous helicopter. In: Proceedings of the International Conference on Robotics and Automation, vol. 1, pp. 558–563 (2003)
18. Hygounenc, E., Jung, I.-K., Soueres, P., Lacroix, S.: The autonomous blimp project of LAAS-CNRS: achievements in flight control and terrain mapping. Int. J. Rob. Res. **23**(4-5), 473–511 (2004)
19. Kim, J., Sukkarieh, S.: Autonomous airborne navigation in unknown terrain environments. IEEE Trans. Aerosp. Electron. Syst. **40**(3), 1031–1045 (2004) July
20. Kwok, N.M., Dissanayake, G.: An efficient multiple hypothesis filter for bearing-only SLAM. In: Proceedings of the 2004 IEEE/RSJ International Conference on Intelligent Robots and Systems, vol. 1, pp. 736–741, October 2004
21. Lacroix, S., Jung, I.K., Soueres, P., Hygounenc, E., Berry, J.P.: The autonomous blimp project of LAAS/CNRS - current status and research challenges. In: Proceeding of the International Conference on Intelligent Robots and Systems, IROS, Workshop WS6 Aerial Robotics, pp. 35–42. IEEE/RSJ (2002)
22. Langedaan, J., Rock, S.: Passive GPS-free navigation of small UAVs. In: Proceedings of the IEEE Aerospace Conference, pp. 1–9 (2005)
23. Lemaire, T., Lacroix, S., Solà, J.: A practical 3D bearing only SLAM algorithm. In: Proceedings of the IEEE/RSJ International Conference on Intelligent Robots and Systems, pp. 2449–2454 (2005)
24. Leonard, J.J., Durrant-Whyte, H.F.: Simultaneous map building and localization for an autonomous mobile robot. In: Proceedings of the IEEE/RSJ International Workshop on Intelligent Robots and Systems, pp. 1442–1447, November 1991
25. Ling, L., Ridley, M., Kim, J.-H., Nettleton, E., Sukkarieh, S.: Six DoF decentralised SLAM. In: Proceedings of the Australasian Conference on Robotics and Automation (2003)
26. Mahony, R., Hamel, T.: Image-based visual servo control of aerial robotic systems using linear image features. IEEE Trans. Robot. **21**(2), 227–239 (2005)
27. Mejías, L., Saripalli, S., Campoy, P., Sukhatme, G.S.: Visual servoing of an autonomous helicopter in urban areas using feature tracking. J. Field. Robot. **23**(3–4), 185–199 (2006)

28. Montiel, J., Civera J, Davison, A.: Unified inverse depth parametrization for monocular SLAM. In: Robotics: Science and Systems, August 2006
29. Ollero, A., Ferruz, J., Caballero, F., Hurtado, S., Merino, L.: Motion compensation and object detection for autonomous helicopter visual navigation in the COMETS system. In: Proceedings of the International Conference on Robotics and Automation, ICRA, pp. 19–24. IEEE (2004)
30. Ollero, A., Merino, L.: Control and perception techniques for aerial robotics. Annu. Rev. Control, Elsevier (Francia), **28**, 167–178 (2004)
31. Papadopoulo, T., Lourakis, M.I.A.: Estimating the jacobian of the singular value decomposition: theory and applications. In: Proceedings of the 2000 European Conference on Computer Vision, vol. 1, pp. 554–570 (2000)
32. Proctor, A.A., Johnson, E.N., Apker, T.B.: Vision-only control and guidance for aircraft. J. Field. Robot. **23**(10), 863–890 (2006)
33. Remuss, V., Musial, M., Hommel, G.: Marvin - an autonomous flying robot-bases on mass market. In: International Conference on Intelligent Robots and Systems, IROS. Proceedings of the Workshop WS6 Aerial Robotics, pp. 23–28. IEEE/RSJ (2002)
34. Saripalli, S., Montgomery, J.F., Sukhatme, G.S.: Visually guided landing of an unmanned aerial vehicle. IEEE Trans. Robot. Autom. **19**(3), 371–380 (2003) June
35. Saripalli, S., Sukhatme, G.S.: Landing on a mobile target using an autonomous helicopter. In: Proceedings of the International Conference on Field and Service Robotics, FSR, July 2003
36. Shakernia, O., Vidal, R., Sharp, C., Ma, Y., Sastry, S.: Multiple view motion estimation and control for landing an aerial vehicle. In: Proceedings of the International Conference on Robotics and Automation, ICRA, vol. 3, pp. 2793–2798. IEEE, May 2002
37. Srinivasan, M.V., Zhang, S.W., Garrant, M.A.: Landing strategies in honeybees, and applications to UAVs. In: Springer Tracts in Advanced Robotics, pp. 373–384. Springer-Verlag, Berlin (2003)
38. Triggs, B.: Autocalibration from planar scenes. In: Proceedings of the 5th European Conference on Computer Vision, ECCV, vol. 1, pp. 89–105. Springer-Verlag, London, UK (1998)
39. Tsai, R.Y., Huang, T.S., Zhu, W.-L.: Estimating three-dimensional motion parameters of a rigid planar patch, ii: singular value decomposition. IEEE Trans. Acoust. Speech Signal Process. **30**(4), 525–534 (1982) August
40. Vidal, R., Sastry, S., Kim, J., Shakernia, O., Shim, D.: The Berkeley aerial robot project (BEAR). In: Proceeding of the International Conference on Intelligent Robots and Systems, IROS, pp. 1–10. IEEE/RSJ (2002)
41. Vidal Calleja, T., Bryson, M., Sukkarieh, S., Sanfeliu, A., Andrade-Cetto, J.: On the observability of bearing-only SLAM. In: Proceedings of the 2007 IEEE International Conference on Robotics and Automation, pp. 1050–4729, April 2007
42. Volpe, J.A.: Vulnerability assessment of the transportation infrastructure relying on the global positioning system. Technical report, Office of the Assistant Secretary for Transportation Policy, August (2001)
43. Wu, A.D., Johnson, E.N., Proctor, A.A.: Vision-aided inertial navigation for flight control. In: Proc. of AIAA Guidance, Navigation, and Control Conference and Exhibit (2005)
44. Yakimenko, O.A., Kaminer, I.I., Lentz, W.J., Ghyzel, P.A.: Unmanned aircraft navigation for shipboard landing using infrared vision. IEEE Trans. Aerosp. Electron. Syst. **38**(4), 1181–1200 (2002) October
45. Zhang, Z., Hintz, K.J.: Evolving neural networks for video attitude and hight sensor. In: Proc. of the SPIE International Symposium on Aerospace/Defense Sensing and Control, vol. 2484, pp. 383–393 (1995) April

Real-time Implementation and Validation of a New Hierarchical Path Planning Scheme of UAVs via Hardware-in-the-Loop Simulation

Dongwon Jung · Jayant Ratti · Panagiotis Tsiotras

Originally published in the Journal of Intelligent and Robotic Systems, Volume 54, Nos 1–3, 163–181.
© Springer Science + Business Media B.V. 2008

Abstract We present a real-time hardware-in-the-loop simulation environment for the validation of a new hierarchical path planning and control algorithm for a small fixed-wing unmanned aerial vehicle (UAV). The complete control algorithm is validated through on-board, real-time implementation on a small autopilot having limited computational resources. We present two distinct real-time software frameworks for implementing the overall control architecture, including path planning, path smoothing, and path following. We emphasize, in particular, the use of a real-time kernel, which is shown to be an effective and robust way to accomplish real-time operation of small UAVs under non-trivial scenarios. By seamless integration of the whole control hierarchy using the real-time kernel, we demonstrate the soundness of the approach. The UAV equipped with a small autopilot, despite its limited computational resources, manages to accomplish sophisticated unsupervised navigation to the target, while autonomously avoiding obstacles.

Keywords Path planning and control · Hardware-in-the-loop simulation (HILS) · UAV

1 Introduction

Autonomous, unmanned ground, sea, and air vehicles have become indispensable both in the civilian and military sectors. Current military operations, in particular,

D. Jung (✉) · J. Ratti · P. Tsiotras
Georgia Institute of Technology, Atlanta, GA 30332-0150, USA
e-mail: dongwon.jung@gatech.edu

J. Ratti
e-mail: jayantratti@gatech.edu

P. Tsiotras
e-mail: tsiotras@gatech.edu

K. P. Valavanis et al. (eds.), *Unmanned Aircraft Systems*. DOI: 10.1007/978-1-4020-9137-7_10   163

depend on a diverse fleet of unmanned (primarily aerial) vehicles that provide constant and persistent monitoring, surveillance, communications, and–in some cases–even weapon delivery. This trend will continue, as new paradigms for their use are being proposed by military planners. Unmanned vehicles are also used extensively in civilian applications, such as law enforcement, humanitarian missions, natural disaster relief efforts, etc.

During the past decade, in particular, there has been an explosion of research related to the control of small unmanned aerial vehicles (UAVs). The major part of this work has been conducted in academia [3–5, 10, 13–15, 23–25], since these platforms offer an excellent avenue for students to be involved in the design and testing of sophisticated navigation and guidance algorithms [17].

The operation of small-scale UAVs brings about new challenges that are absent in their large-scale counterparts. For instance, *autonomous* operation of small-scale UAVs requires both trajectory design (planning) and trajectory tracking (control) tasks to be completely automated. Given the short response time scales of these vehicles, these are challenging tasks using existing route optimizers. On-board, real-time path planning is especially challenging for small UAVs, which may not have the on-board computational capabilities (CPU and memory) to implement some of the sophisticated path planning algorithms proposed in the literature. In fact, the effect of limited computational resources on the control design of real-time, embedded systems has only recently received some attention in the literature [1, 29]. The problem is exacerbated when a low-cost micro-controller is utilized as an embedded control computer.

Autonomous path planning and control for small UAVs imposes severe restrictions on control algorithm development, stemming from the limitations imposed by the on-board hardware and the requirement for real-time implementation. In order to overcome these limitations it is imperative to develop computationally efficient algorithms that make use of the on-board computational resources wisely.

Due to the stringent operational requirements and the hardware restrictions imposed on the small UAVs, a complete solution to fully automated, unsupervised, path planning and control of UAVs remains a difficult undertaking. Hierarchical structures have been successfully applied in many cases in order to deal with the issue of complexity. In such hierarchical structures the entire control problem is subdivided into a set of smaller sub-control tasks (see Fig. 1). This allows for a more straightforward design of the control algorithms for each modular control task. It also leads to simple and effective implementation in practice [2, 27, 28].

In this paper, a complete solution to the hierarchical path planning and control algorithm, recently developed by the authors in Refs. [16, 21, 30], is experimentally validated on a small-size UAV autopilot. The control hierarchy consists of path planning, path smoothing, and path following tasks. Each stage provides the necessary commands to the next control stage in order to accomplish the goal of the mission, specified at the top level. The execution of the entire control algorithm is demonstrated through a realistic hardware-in-the-loop (HIL) simulation environment. All control algorithms are coded on a micro-controller running a real-time kernel, which schedules each task efficiently, by taking full advantage of the provided kernel services. We describe the practical issues associated with the implementation of the proposed control algorithm, while taking into consideration the actual hardware limitations.

We emphasize the use of a real-time kernel for implementing the overall control architecture. A real-time operating system provides the user with great flexibility in building complex real-time applications [11], owing to the ease in programming, error-free coding, and execution robustness. We note in passing that currently there exist many real-time kernels employed for real-time operation of UAVs. They differ in the kernel size, memory requirements, kernel services, etc. Some of these real-time kernels can be adopted for small micro-controller/processor [6, 12]. An open source real-time Linux is used for flight testing for UAVs [8]. In this work we have used the MicroC/OS-II [26], which is ideal for the small microcontroller of our autopilot.

2 Hierarchical Path Planning and Control Algorithm

In this section, we briefly describe a hierarchical path planning and control algorithm, which has been recently developed by the authors, and which takes into account the limited computational resources of the on-board autopilot.

Figure 1 shows the overall control hierarchy. It consists of path planning, path smoothing, path following, and the low level autopilot functions. At the top level of the control hierarchy, a wavelet-based, multiresolution path planning algorithm [21, 30] is employed to compute an optimal path from the current position of

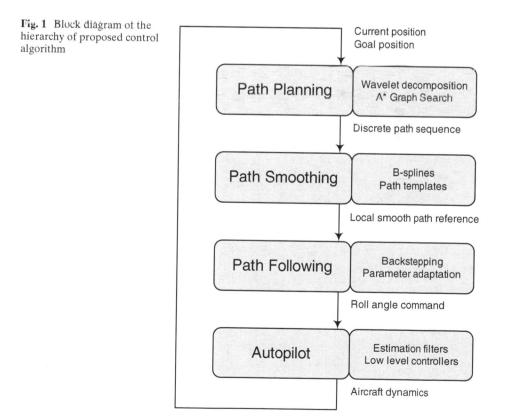

Fig. 1 Block diagram of the hierarchy of proposed control algorithm

the vehicle to the goal. The path planning algorithm utilizes a multiresolution decomposition of the environment, such that a coarser resolution is used far away from the agent, whereas fine resolution is used in the vicinity of the agent. The result is a topological graph of the environment of known a priori complexity. The algorithm then computes a path with the highest accuracy at the current location of the vehicle, where is needed most. Figure 2 illustrates this idea. In conjunction with the adjacency relationship derived directly from the wavelet coefficients [21], a discrete cell sequence (i.e., channel) to the goal destination is generated by invoking the $\mathcal{A}^*$ graph search algorithm [7, 9].

The discrete path sequence is subsequently utilized by the on-line path smoothing layer to generate a smooth reference path, which incorporates path templates comprised of a set of B-spline curves [22]. The path templates are obtained from an off-line optimization step, so that the resulting path is ensured to stay inside the prescribed cell channel. The on-line implementation of the path smoothing algorithm finds the corresponding path segments over a finite planning horizon with respect to the current position of the agent, and stitches them together, while preserving the smoothness of the composite curve.

After a local smooth reference path is obtained, a nonlinear path following control algorithm [20] is applied to asymptotically follow the reference path constructed by the path smoothing step. Assuming that the air speed and the altitude of the UAV are constant, a kinematic model is utilized to design a control law to command heading rate. Subsequently, a roll command to follow the desired heading rate is computed by taking into account the inaccurate system time constant. Finally, an autopilot with on-board sensors that provides feedback control for the attitude angles, air speed, and altitude, implements the low-level inner loops for command following to attain the required roll angle steering, while keeping the altitude and the air speed constant.

As shown in Fig. 1, at each stage of the hierarchy, the corresponding control commands are obtained from the output of the previous stage, given the initial

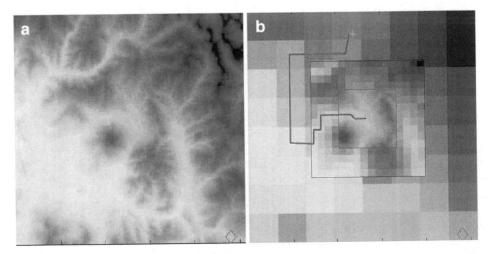

Fig. 2 Demonstration of multi-resolution decomposition of the environment (**a**) using square cells induced by the use of Haar wavelets (**b**). The current location of the agent (vehicle) is at the center of the red square (high-resolution region). The dynamics are included in the high-resolution region

environment information (e.g., a two dimensional elevation map). With the goal position specified by the user, this hierarchical control algorithm allows the vehicle to accomplish its mission of reaching the goal destination, while avoiding obstacles.

3 Experimental Test-Bed

3.1 Hardware Description

A UAV platform based on the airframe of an off-the-shelf R/C model airplane has been developed to implement the hierarchical path planning and control algorithm described above. The development of the hardware and software was done completely in-house. The on-board autopilot is equipped with a micro-controller, sensors and actuators, and communication devices that allow full functionality for autonomous control. The on board sensors include angular rate sensors for all three axes, accelerometers along all three axes, a three-axis magnetic compass, a GPS sensor, and absolute and differential pressure sensors. An 8-bit micro-controller (Rabbit RCM-3400 running at 30 MHz with 512 KB RAM and 512 KB Flash ROM) is the core of the autopilot. The Rabbit RCM-3400 is a low-end micro-controller with limited computational throughput (as low as 7 μs for floating-point multiplication and 20 μs for computing a square root) compared to a generic high performance 32 bit micro-processor. This micro-controller provides data acquisition, data processing, and manages the communication with the ground station. It also runs the estimation algorithms for attitude and absolute position and the low-level control loops for the attitude angles, air speed, and altitude control. A detailed description of the UAV platform and the autopilot can be found in Refs. [17, 18].

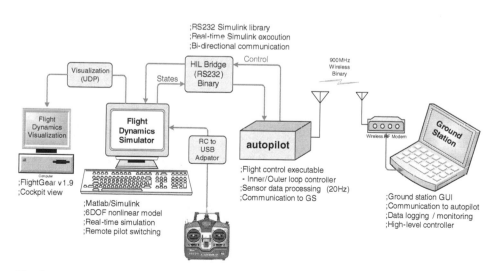

Fig. 3 High fidelity hardware-in-the-loop simulation (HILS) environment for rapid testing of the path planning and control algorithm

3.2 Hardware-in-the-Loop Simulation Environment

A realistic hardware-in-the-loop simulation (HILS) environment has been developed to validate the UAV autopilot hardware and software operations utilizing Matlab® and Simulink®. A full 6-DOF nonlinear aircraft model is used in conjunction with a linear approximation of the aerodynamic forces and moments, along with gravitational (WGS-84) and magnetic field models for the Earth. Detailed models of the sensors and actuators have also been incorporated. Four independent computers are used in the HILS, as illustrated in Fig. 3. A 6-DOF simulator, a flight visualization computer, the autopilot micro-controller, and the ground station computer console are involved in the HIL simulation. Further details about the HILS set-up can be found in Ref. [19].

4 Real-Time Software Environment

The software architecture of the on-board autopilot is shown in Fig. 4. It is comprised of several blocks, called *tasks*, which are allotted throughout different functioning layers such as the application level, the low level control, the data processing level, and the hardware level. The tasks at the hardware level, or hardware services, interact with the actual hardware devices to collect data from the sensors, communicate with the ground station, and issue commands to the DC servo motors. The

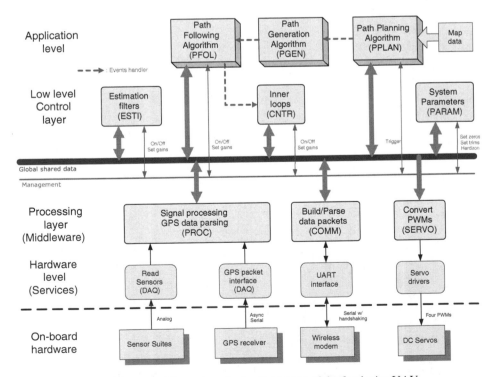

Fig. 4 Software architecture of the on-board autopilot system of the fixed-wing UAV

middleware tasks on top of the hardware services provide the abstraction of the inbound and outbound data, by supplying the processed data on a globally shared data bus or by extracting data from the global bus to the corresponding hardware services. Utilizing the processed data on the globally shared data bus, the lower level control layer achieves the basic control functions such as estimation of the attitude angles, estimation of the absolute position, and implementation of the inner loop PID controllers. Finally, three application tasks, which correspond to path planning, path generation, and path following, are incorporated to implement the hierarchical control architecture described in Section 2. The hierarchical control structure dictates all application tasks, in the sense that completion of the upper level task (event triggered) initiates the execution of a lower level task (event processed). This is shown by red dashed arrows in Fig. 4 each representing an event signal. In Fig. 4, besides exchanging the data via the global shared data bus, each task is managed by a global management bus, used for triggering execution of tasks, initializing/modifying system parameters, etc.

The task management, also called task scheduling, is the most crucial component of a real-time system. It seamlessly integrates the multiple "tasks" in this real-time software application. In practice however, a processor can only execute one instruction at a time; thus multitasking scheduling is necessitated for embedded control system implementations where several tasks need to be executed while meeting real-time constraints. Using multitasking, more than one task, such as control algorithm implementation, hardware device interfaces and so on, can appear to be executed in parallel. However, the tasks need to be prioritized based on their importance in the software flow structure so that the multitasking kernel correctly times their order of operation, while limiting any deadlocks or priority inversions.

4.1 Cooperative Scheduling Method: Initial Design

For the initial implementation, we developed a near real-time control software environment that is based predominately on the idea of cooperative scheduling. Cooperative scheduling is better explained by a large main loop containing small fragments of codes (tasks). Each task is configured to voluntarily relinquish the CPU when it is waiting, allowing other tasks to execute. This way, one big loop can execute several tasks in parallel, while no single task is busy waiting.

Like most real-time control problems, we let the main loop begin while waiting for a trigger signal from a timer, as shown by the red arrows in Fig. 5. In accordance with the software framework of Fig. 4, we classify the tasks into three groups: *routine tasks, application tasks*, and *non-periodic tasks*. The routine tasks are critical tasks required for the UAV to perform minimum automatic control. In our case these consist of the tasks of reading analog/GPS sensors (DAQ), signal processing (PROC), estimation (ESTI), inner loop control (CNTR), and servo driving (SERVO). The sampling period T_s is carefully chosen to ensure the completion of the routine tasks and allow the minimum sampling period to capture the fastest dynamics of the system. In order to attain real-time scheduling over all other tasks besides the routine tasks, a sampling period of $T_s = 50$ ms, or a sampling rate of 20 Hz, was used. On the other hand, some of the application tasks require substantial computation time and resources, as they deal with the more complicated, high level computational algorithms such as path planning (PPLAN), path generation (PGEN), and path

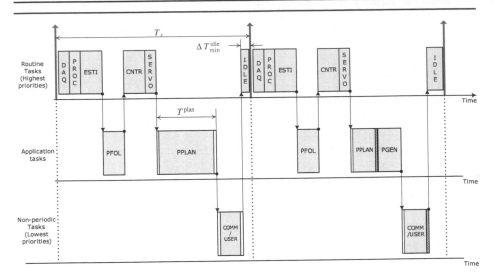

Fig. 5 A real-time scheduling method combining cooperative and naive preemptive multitasking

following (**PFOL**). In particular, the path planning algorithm in Ref. [21] turns out to have a total computation time greater than the chosen sampling period. As a result, and in order to meet the real-time constraints, we fragmentized the execution of the computationally intensive task, **PPLAN** into several slices of code execution, each with a finite execution time T^{plan}. The finite execution time is selected a priori by taking into account both T_s and the (estimated) total execution time of the routine tasks. The objective is to maximize the CPU usage to complete the task **PPLAN** as soon as possible, while meeting the criterion for real-time operation. Finally, non-periodic tasks such as communication (**COMM**) and user application (**USER**) are executed whenever the CPU becomes available, ensuring a minimum idling time of duration $\Delta T_{\mathrm{min}}^{\mathrm{idle}}$ to allow the CPU to wait for other triggering signals.

Figure 6 shows a pseudo-code implementation of the cooperative scheduling scheme. Each `costate` implements the cooperative scheduling, while the `slice` statement implements the naive preemptive scheduling, which preempts the CPU over the finite execution window T^{plan}.

4.2 Preemptive Scheduling Method: Final Design

Given the a priori knowledge of the required tasks to be executed, in conjunction with an approximate knowledge of the total execution time, the use of *costate* blocks was shown to be an effective implementation of cooperative scheduling in the previous section. However, it is often a painstaking job for a programmer to meticulously synchronize and schedule all tasks in the application, many of which may have unpredictable execution time. Alternatively, it is possible to design a cooperative scheduler using conservative timing estimates for the corresponding tasks in a manner similar to that of Section 4.1. However, such an approach will result in poor performance in terms of the overall completion time. With a conservative estimate of execution times for the routine tasks, the portion allotted for the

Fig. 6 Pseudo-code implementation of the combined cooperative/preemptive scheduling scheme for the hierarchical path planning and control algorithm

```
main() {
    while (1) {
        costate {
            Wait_for_timer (Tₛ);
            Task DAQ;
            Task PROC;
            Task ESTI;
            if (EVENT(PFOL)) Task CNTR;
            Task SERVO;
        }
        costate {
            if (EVENT(PGEN)) Task PFOL;
        }
        costate {
            if (EVENT(PPLAN)) Task PGEN;
        }
        costate {
            Task COMM;
            Task PARAM;
            Task USER;
        }
        if (ΔTⁱᵈˡᵉ > ΔTᵖˡᵃⁿ) {
            slice (ΔTᵖˡᵃⁿ, Task PPLAN);
        }
    }
}
```

execution of the computationally expensive tasks remains fixed regardless whether the CPU remains idle for the rest of the sampling period. This implies that the CPU does not make full use of its capacity, thus delaying the execution of the overall tasks by a noticeable amount of time. The throughput of the computationally intensive tasks may be improved by employing a *preemptive multitasking scheduler* [26]. Since the kernel has full access to CPU timing, it can allot the CPU resources to the lower level tasks whenever the higher level tasks relinquish their control. This effectively minimizes the CPU idle time and reduces the task completion time. In the next section we present an alternative framework for implementing the hierarchical path planning and control algorithm shown in Section 2 using a preemptive real-time kernel, namely, MicroC/OS-II.

The MicroC/OS-II is known to be a highly portable, low on memory, scalable, preemptive, real-time operating system for small microcontrollers [26]. Besides being a preemptive task scheduler which can manage up to 64 tasks, the MicroC/OS-II also provides general kernel services such as semaphores, including mutual exclusion semaphores, event flags, message mailboxes, etc. These services are especially helpful for a programmer to build a complex real-time software system and integrate tasks seamlessly. Its use also simplifies the software structure by utilizing a state flow concept. The MicroC/OS-II allows small on-chip code size of the real-time kernel. The code size of the kernel is no more than approximately 5 to 10 kBytes [26], adding a relatively small overhead (around 5.26%) to the current total code size of 190 kBytes for our application.

4.2.1 Real-time software architecture

Real-time software programming begins with creating a list of tasks. In this work we emphasize the real-time implementation of the path planning and control algorithm using HILS. This requires new tasks to deal with the additional HILS communication. The simulator transmits the emulated sensor data to the micro-controller via serial communication. Hence, the sensor/GPS reading task (DAQ) is substituted with the sensor data reading task (HILS_Rx), which continuously checks a serial buffer for incoming communication packets. Similarly, the servo driving task (SERVO) is replaced by the command writing task (HILS_Tx), which sends back PWM servo commands to the simulator. On the other hand, the communication task COMM is subdivided into three different tasks according to their respective roles, such as a downlink task for data logging (COMM_Tx), an uplink task for user command (COMM_Rx), and a user command parsing task (COMM_Proc). In addition, we create a path management (PMAN) task which coordinates the execution of the path planning and control algorithm, thus directly communicating with PPLAN, PGEN, and PFOL, respectively. Finally, a run-time statistics checking task (STAT) is created in order to obtain run-time statistics of the program such as CPU usage and the execution time of each task. These can be used to benchmark the performance of the real-time kernel. Table 1 lists all tasks created in the real-time kernel.

The MicroC/OS-II manages up to 64 distinct tasks, the priorities of which must be uniquely assigned. Starting from zero, increasing numbers impose lower priorities to be assigned to corresponding tasks. In particular, because the top and bottom ends of the priority list are reserved for internal kernel use, application tasks are required to have priorities other than a priority level in this protected range. Following an empirical convention of priority assignment, we assign the critical tasks with high priorities because they usually involve direct hardware interface. In order to minimize degradation of the overall performance of the system, the hardware

Table 1 List of tasks created by the real-time kernel

ID	Alias	Description	Priority
1	HILS_Tx	Sending back servo commands to the simulator	11
2	HILS_Rx	Reading sensor/GPS packets from the simulator	12
3	COMM_Rx	Uplink for user command from the ground station	13
4	COMM_Proc	Parsing the user command	14
5	ESTI_Atti	Attitude estimation task	15
6	ESTI_Nav	Absolute position estimation task	16
7	CNTR	Inner loop control task	17
8	PFOL	Nonlinear path following control task	18
9	COMM_Tx	Downlink to the ground station	19
10	PGEN	Path generation task using B-spline templates	20
11	PMAN	Control coordination task	21
13	PPLAN	Multiresolution path planning task	23
12	STAT	Obtaining run-time statistics	22

related tasks may need proper synchronization with the hardware, hence demanding immediate attention. It follows that routine tasks that are required for the UAV to perform minimum automatic control such as ESTI_Atti, ESTI_Nav, and CNTR are given lower priorities. Finally, application-specific tasks such as PFOL, PGEN, and PPLAN are given even lower priorities. This implies that these tasks can be activated whenever the highest priority tasks relinquish the CPU. Table 1 shows the assigned priority for each task. Note that the task COMM_Tx is assigned with a lower priority because this task is less critical to the autonomous operation of the UAV.

Having the required tasks created, we proceed to design a real-time software framework by establishing the relationships between the tasks using the available kernel services: A semaphore is utilized to control access to a globally shared object in order to prevent it from being shared indiscriminately by several different tasks. Event flags are used when a task needs to be synchronized with the occurrence of multiple events or relevant tasks. For inter-task communication, a mailbox is employed to exchange a message in order to convey information between tasks.

Figure 7 illustrates the overall real-time software architecture for the autopilot. In the diagram two binary semaphores are utilized for two different objects corresponding to the wireless modem and a reference path curve, respectively. Any task that requires getting access to those objects needs to be blocked (by semaphore pending) until the corresponding semaphore is either non-zero or is released (by semaphore posting). Consequently, only one task has exclusive access to the objects at a time, which allows data compatibility among different tasks. The event flags are posted by the triggering tasks and are consumed by the pending tasks, allowing synchronization of two consecutive tasks. Note that an event from the HILS_Rx triggers a chain of routine tasks for processing raw sensor data (ESTI_Atti, ESTI_Nav) and control implementation (CNTR). A global data storage is used to hold all significant variables that can be referenced by any task, while the global flags declaration block contains a number of event flag groups used for synchronization of tasks. Each mailbox can hold a byte-length message which is posted by a sender task with information indicated next to the data flow arrow symbol. Each task receiving the message will empty the mailbox and wait for another message to be posted. These mailboxes are employed to pass the results of one task to another task. It should be noted that when the task PMAN triggers the task PPLAN, the results of subsequent tasks are transmitted via mailboxes in the following order: (PPLAN → PGEN → PMAN → PFOL).

4.2.2 Benefits of using a real-time kernel

Robustness: The real-time kernel provides many error handling capabilities during deadlock situations. We have been able to resolve all possible deadlocks using the timing features of the Semaphore-Pend or Flag-Pend operations. The kernel provides time-out signals in the semaphore and flag calls with appropriate generated errors. These are used to handle the unexpected latency or deadlock in the scheduling operation.

Flexibility and ease of maintenance: The entire architecture for the autopilot software has been designed, while keeping in mind the object-oriented requirements for an applications engineer. The real-time kernel provides an easy and natural way to achieve this goal. The architecture has been designed to keep the code flexible enough to allow adding higher level tasks, like the ones required to process

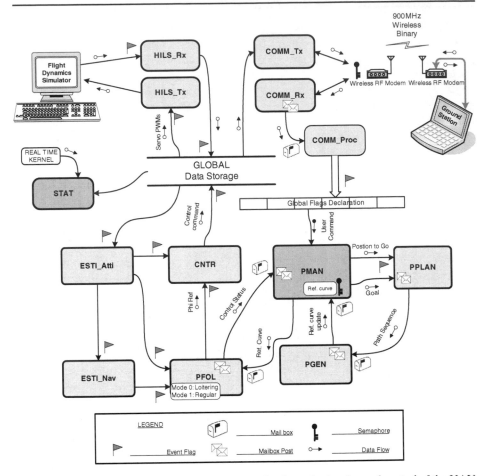

Fig. 7 The complete real-time software architecture for the path planning and control of the UAV

and execute the multi-resolution wavelet path planning algorithm. All this can be achieved without engrossing into the system level intricacies of handling and programming a microcontroller/microprocessor. The flexibility of the architecture also makes it extremely efficient to debug faults in low-level, mid-level or high-level tasks, without having to re-code/interfere with the other tasks.

5 Hardware-in-the-Loop Simulation Results

In this section we present the results of the hierarchical path control algorithm using a small micro-controller in a real-time HILS environment. The details of the implementation are discussed in the sequel.

5.1 Simulation Scenario

The environment is a square area containing actual elevation data of a US state, of dimension 128×128 units, which corresponds to 9.6×9.6 km. Taking into account the available memory of the micro-controller, we choose the range and granularity of the fine and coarse resolution levels. Cells at the fine resolution have dimensions 150×150 m, which is larger than the minimum turning radius of the UAV. The minimum turn radius is approximately calculated for the vehicle flying at a constant speed of $V_T = 20$ m/s with a bounded roll angle of $|\phi| \leq 30°$, resulting in a minimum turn radius of $R_{min} \approx 70$ m.

The objective of the UAV is to generate and track a path from the initial position to the final position while circumventing all obstacles above a certain elevation threshold. Figure 8 illustrates the detailed implementation of the proposed path planning and control algorithm. Initially, the UAV is loitering around the initial position p_0 until a local path segment from p_0 to p_a is computed (Step A,B). Subsequently, the path following controller is engaged to follow the path (Step C,D). At step D, the UAV replans to compute a new path from the intermediate location

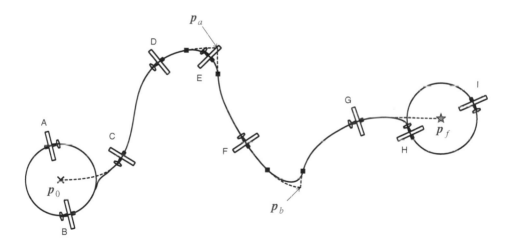

Step	Task description
A	Initially, the UAV is loitering around the initial position with the circle radius R_l
B	Calculate the first path segment from p_0 to p_a
C	Break away from the loitering circle, start to follow the first path segment
D	Calculate the second segment from p_a to p_b, and a transient path
E	UAV is on the transient path
F	Calculate the third path segment, and a transient path
G	UAV is approaching the goal position, no path is calculated
H	The goal is reached, end of the path control, get on the loitering circle
I	UAV is loitering around the goal position p_f

Fig. 8 Illustration of the real-time implementation of the proposed hierarchical path planning and control algorithm

Fig. 9 HILS results of the hierarchical path planning and control implementation. The *plots on the right* show the close-up view of the simulation. At each instant, the channel where the smooth path segment from the corresponding path template has to stay, is drawn by polygonal lines. The actual path followed by the UAV is drawn on top of the reference path. **a** $t = 64.5$ s. **b** $t = 333.0$ s. **c** $t = 429.0$ s. **d** $t = 591.5$ s

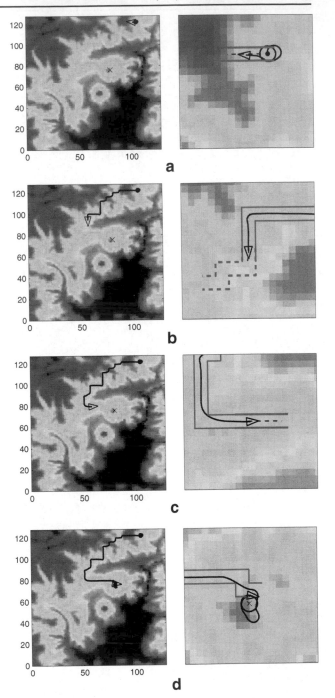

$\boldsymbol{p}_a$ to the goal, resulting in the second local path segments from $\boldsymbol{p}_a$ to $\boldsymbol{p}_b$. The first and second path segments are stitched by a transient B-spline path assuring the continuity condition at each intersection point (marked by black squares). This process iterates